# *Algonquin* WILDLIFE
*Lessons in Survival*

## NORM QUINN
### Foreword by Dan Strickland

Original drawings by Cassandra Ward

NATURAL HERITAGE BOOKS

Algonquin Wildlife: Lessons in Survival
Norm Quinn

Copyright © 2002 Norm Quinn

All rights reserved. No portion of this book, with the exception of brief extracts for the purpose of literary or scholarly review, may be reproduced in any form without the permission of the publisher.

Published by Natural Heritage / Natural History Inc.
P.O. Box 95, Station O, Toronto, Ontario M4A 2M8

www.naturalheritagebooks.com

National Library of Canada Cataloguing in Publication

Quinn, Norm, 1951-
   Algonquin wildlife : lessons in survival / Norm Quinn.

Includes bibliographical references and index.
ISBN 1-896219-28-4

1. Wildlife research—Ontario—Algonquin Provincial Park—History.
2. Animals—Ontario—Algonquin Provincial Park. 3. Algonquin Provincial Park (Ont.)—History. I. Title.

QL84.26.O5Q84 2002     591.9'713'147     C2002-902774-8

Cover and text design by Sari Naworynski
Edited by Jane Gibson
Printed and bound in Canada by Hignell Printing Limited, Winnipeg, Manitoba
All black-and-white drawings © Cassandra Ward

Front cover visuals: Moose, *Algonquin Park Museum Archives, M.N.R. Staff*; Grey Jay, Snapping Turtles, *Norm Quinn*.

Back cover visuals: Prof. D. Pimlott, *Algonquin Park Museum Archives – Mark Pimlott*; Deer, *Algonquin Park Museum Archives #6196 – John R. Needham*.

Spine: Bear, *Norm Quinn*.

Natural Heritage / Natural History Inc. acknowledges the financial support of the Canada Council for the Arts and the Ontario Arts Council for our publishing program. We also acknowledge the financial support of the Government of Canada through the Book Publishing Industry Development Program (BPIDP) and the Association for the Export of Canadian Books.

*To Nancy, Robert and Laura*

*Edward "Ned" Godin, an early park ranger, retired in 1932. Ned was stationed mainly at Achray and is identified as the builder of Out-Side Inn, the ranger cabin that still stands there, now housing exhibits. He was also known to have been a fire-ranging partner and a close friend of Tom Thomson, who used to board with him while painting at Algonquin.[1]*
Ontario Ministry of Natural Resources, Publisher's Collection.

# Table of Contents

| | |
|---|---|
| Foreword by Dan Strickland | *vi* |
| Acknowledgements | *viii* |
| Author's Note | *viii* |
| Preface | *ix* |
| 1 │ The Quest | *1* |
| 2 │ Some Davids and a Goliath | *6* |
| 3 │ Hemlock and History | *23* |
| 4 │ Of Tooth and Claw | *47* |
| 5 │ And Hares … and Bears | *85* |
| 6 │ Of Time and Trout | *101* |
| 7 │ Stress – Vulnerability Abounds | *120* |
| 8 │ The Twig Eaters | *136* |
| 9 │ Moose Days and Jays | *159* |
| 10 │ Bet Hedging | *186* |
| 11 │ The Station | *204* |
| Appendix I – Technical Expansion on Selected Topics | *214* |
| Notes | *219* |
| Bibliography | *237* |
| Index | *238* |
| About the Author | *246* |
| About the Illustrator | *246* |

# Foreword

by Dan Strickland

Algonquin and wildlife are synonymous in the minds of many people. In its early years, Ontario's first and foremost provincial park was *the* place in the province to see and feed white-tailed deer. Later, in the 1980s and '90s, Algonquin became equally renowned as the best place in Ontario to observe moose. And, most famous of all, the Park is celebrated today as the place where more people have had first-hand contact with wolves than anywhere else in the world.

Nor do Algonquin's wildlife riches end there. With its mix of northern spruce and southern hardwoods and pines, the Park is simultaneously the home of northern birds and mammals that one could find on the shores of James Bay and also of southern species typical of Lake Erie country. Indeed, from a wildlife point of view, one would be hard pressed to find a single place more representative of Ontario as a whole than Algonquin. This diversity has long been recognized by southern Ontario birders, most of whom make an annual pilgrimage to Algonquin to see northern specialties such as Grey Jays and Spruce Grouse. The Park is also an automatic destination of overseas ecotourists visiting Canada and seeking to see as much of our wildlife as possible in just a few places and under the best possible viewing conditions.

It is perhaps less well known that Algonquin has also been a magnet for wildlife biologists, the people who study our country's birds and mammals and try to learn how they live and interact with each other and their environment. The Park has records of wildlife research undertaken in Algonquin as early as the 1920s and '30s but things really took off when the Wildlife Research Station was established in the 1940s (at first in an abandoned lumber mill at Lake of Two Rivers and then later at nearby Lake Sasajewun). Today, the number of scientific reports and articles based on research in Algonquin Park has reached over 700 and the explosion shows no sign of slowing down.

Many of the interpretations and insights gained from this work have made their way into Park publications, magazine articles and popular books – and have therefore been assimilated into general public knowledge. Still, it is also true that much has been "lost in translation." Even the popular works – and certainly the scientific articles – utterly fail to portray the personalities of the researchers who have come and gone in Algonquin. Nor, for the most part, do they come anywhere close to conveying any sense of the long and sometimes torturous paths of observation and reasoning that led to the conclusions reached by these personalities.

But fear not! The book you are about to read will go a long way towards making up for these failings. For eighteen years Norm Quinn has been both a witness and a participant in the busy wildlife research community in Algonquin. He is especially well qualified to capture the true flavour of the wonderful intellectual discipline which seeks to understand the rules by which our fellow animals live in our world. And, as you will see, Norm also has an appreciation for the funny stories behind this endeavour, and a real flair for telling them. As a sometime participant and observer of the Park wildlife scene myself, I greatly enjoyed and appreciated this book. I thank Norm for writing it and for preserving some big chapters of Park history in such a humorous manner.

I hope that you, too, will find it as enjoyable and interesting as I did.

Dan Strickland
Chief Park Naturalist at Algonquin Park from 1970 to 2000

# Acknowledgements

I would like to acknowledge all of the biologists who have given me inspiration and in particular, Ron Brooks, Harry Lumsden, Dan Keppie, Jim Gardner, Ed Addison, Dennis Voigt and Dan Strickland. I would also like to acknowledge the dozens of scientists who have conducted the great body of brilliant and fascinating wildlife research in Algonquin Park of which this book is a celebration. Finally, I am also indebted to Cassandra Ward of Bancroft, Ontario, for the line drawings.

# Author's Note

The opinions expressed in this book are those of the Author only and do not represent those of his employer, the Ontario Ministry of Natural Resources. The author has written this book as a private citizen and not as an employee of the Ministry of Natural Resources.

# Preface

I have had the great privilege of being Park Management Biologist in Algonquin Park for seventeen years. During that time I've seen a small army of wildlife researchers come and go, and have had the good fortune of getting to know many of these fascinating people on a personal basis. There has probably been more research done in Algonquin Park than in any other protected landscape in the world; biologists have, at one time or other, studied everything from parasites in the blood of frogs to the rutting behaviour of moose. This book is a celebration of all this work, an attempt to draw some of the most exciting and important studies in wildlife science out from the dry pages of biological journals into a set of entertaining, reader-friendly stories. I have tried to bring together some of the key themes of wildlife biology into a compelling, fast-paced and, I hope, amusing narrative.

This is also a book about Algonquin Park itself. Algonquin is Ontario's flagship Park and the foremost canoe-tripping wilderness in the world. If you are one of the thousands of devotees of the Park you'll find this book opens a window into a relatively unknown but vital part of the Park's heritage. This book is, lastly, about Algonquin's researchers, the dedicated, eccentric and downright quirky personalities that have roamed and probed Algonquin's forests and lakes in pursuit of nature's secrets, big and small. To them I owe a great debt for I am merely their scribe; the stories are theirs.

# 1 | The Quest

*What a dull world if we knew all about geese.*[1]
— Aldo Leopold, 1966.

There is a popular print titled "The Young Biologist" that you can see hanging in any number of living rooms and halls around southern Ontario. The print shows a young boy in circa 1890s dress staring pensively at a frog (a wood frog actually) in his yard.[2] My mother bought me a copy years ago because the print reminded her so much of me. I am, you see, a biologist, and I spent much of my childhood wandering the woods and fields near our home north of Montreal, collecting crickets and snakes and staring spellbound for hours at frogs or ducks or anything else that would sit still long enough. The print hangs prominently in our living room and I never pass by it without a whimsical drift back to that idyllic childhood.

*A Wood Frog.*

I was, to be sure, a strange sort of kid – a "geek" in today's parlance. While other boys were out playing baseball or teasing girls, I was off on my own birdwatching or turning over rocks to see what might crawl out (I still, to this day, dislike baseball). Although I was an oddball, and knew it, it was an idyllic time and "The Young Biologist" never fails to evoke a warm feeling for those days of wonder.

The print, though, produces much more in me than mere nostalgia. To me, the young boy's fascination with the frog symbolizes man's deep-seated need to understand nature, to comprehend the workings of wild things. I believe that this great longing to understand nature and in particular the variations and cycles of wildlife is deeply rooted in our primitive psyche and has, in modern times, given birth to an entirely new profession. That profession is my profession: wildlife biology.

Let me explain.

The modern science of wildlife biology got its start just after the end of World War II. At the time, foresters, who were already practising a well-developed science, began to broaden their thinking to, in essence, "see the forest for the trees." Some of these innovators, most notably a visionary by the name of Aldo Leopold, left forestry and began to lay the groundwork for the science of wildlife management. Leopold wrote a series of essays, most notably *A Sand County Almanac*,[3] that laid the foundation of the new profession. At about the same time, a group of serious scientists, men like Paul Errington who studied mink predation on muskrats, emerged to make the first real efforts at wildlife research.

From the start, the emphasis has been on one very basic question: *what determines the abundance of animals in nature?* Practically all wildlife work is in some way or another connected to that one fundamental problem. Biologists spend most of their time working with the balance sheet of life and death; tallying the living and dead in an attempt to explain the frequent, sometimes dramatic, and often baffling fluctuations of wildlife populations. The "balance of nature" is, as we shall see, a myth; nothing is so basic to nature as change, and the purpose, the great quest, of wildlife science, has been to understand such changes. Why, for example, are deer so abundant in North

America right now? Why is the enigmatic little woodcock, the "timberdoodle" of our north woods, declining? Why are peregrine falcons, once endangered, now recovering?

Actually, the first serious look at animal numbers was done by a couple of entomologists (insect specialists). Two Australians, Herbert Andrewartha and Louis Birch did a monumental study of insects in the 1950s and published a grinding 800-page tome on their efforts, *The Distribution and Abundance of Animals*,[4] that became a classic work of science. Wildlifers took inspiration from the relentless Aussies and applied their work with bugs to bigger things.

I have often thought that this longing to understand nature must be an ancient obsession, a mirror on our distant past. We are, after all, descended from hunter-gatherers, people with a life-and-death dependence on wildlife. Primitive man must have puzzled (and at times anguished) over cycles of wildlife abundance.

Not long ago I visited the famous Lascaux caves in France, the site of the best known Paleolithic cave paintings.[5] The images are stunning – hundreds of beautifully detailed paintings of oxen, stags, reindeer and now-extinct rhinos, all done in colours that remain vivid and alive to this day. What really struck me, though, was that almost all of the images are of game animals; food on the hoof. These people were portraying with religious reverence what must have been the greatest anxiety of their day-to-day lives: the cruel unpredictability of their food supply. They must have obsessed over the whys and wherefores of the abundance of animals. My own professional interest was staring back at me from 30,000 years past.

Primitive man, we know, had a deep understanding of the habits of wildlife, particularly those species that he hunted. Native North Americans revered game animals, in some cases considering them to have a higher spiritual rank than humans.[6] Our distant ancestors must have understood the basics of ecology and developed some sort of "science" on the subject; more, perhaps, than we will ever appreciate. If so, my profession may be the most ancient of all, older, even, than that less respectable occupation that is commonly accorded the honour.

There is perhaps nowhere on earth that this quest to understand wildlife has borne fruit more richly than Algonquin Park, Ontario.

*This aerial view of Algonquin Park was taken in 1950 by photographer Quimby Hess.* Ontario Ministry of Natural Resources, Publisher's Collection.

Algonquin is one of the great wild Parks of the world. A 7,700-square-kilometre mosaic of crystal clear lakes and stately maple and pine forests, the Park straddles the rugged highlands of south-central Ontario like a giant verdant saddle. Algonquin contains elements of both the coniferous forests of northern Canada and broad-leaved deciduous forests to the south, thus providing for an exceptionally rich mix of plants and animals. With its over 1,400 lakes, the Park was established with great foresight in the late 1800s to protect these source waters for the farms and settlements developing on its periphery. Today, these lakes are linked by a vast web of hundreds of kilometres of portages – a canoe-tripper's paradise. Travel by canoe is, in fact, the only legal means of entry into much of the Park's isolated central core, making the interior a haven of solitude for city-weary souls.

And the city is an ever-present part of the Park, for Algonquin is an easy drive from much of the industrial heartland of Ontario. Cities contain centres of higher learning, providing a proximity of nearly a dozen universities. This, combined with the diversity of the Park's wildlife and what I believe to be a peculiarly Canadian interest in

*Above: Wagon laden with canoes, on portage in Algonquin Park, circa 1897.* Publisher's Collection.

*Left: Feeding fawns near Cache Lake in 1928.* Algonquin Park Museum Archives #6196 – John R. Needham.

things natural, has made the Park a magnet for researchers. It is said that there has been more research done in Algonquin Park than any protected landscape anywhere, a position supported by the over 1,800 publications in the Park's Bibliography.[7]

With wildlife having the great draw on us that it does, most of these researchers were wildlife biologists and all participated, in one way or other, in the same quest that compelled Leopold, Andrewartha and Birch, the Lascaux painters, and, in my mind's eye, the "Young Biologist." They were, as we shall see, a brilliant, driven and quirky lot, and none more so than the gifted but eccentric group we will look at first.

# 2 | Some Davids and a Goliath

> *The thing-in-itself, the will-to-live, exists whole and undivided in every being, even the tiniest.*[1]
> – Scholpenhauer, 1851.

Deep in a pinewoods in the core of Algonquin Park, carefully hidden from the throng of nearby summer tourists, lies a sprawling complex of musty cabins named rather pretentiously the Wildlife Research Station. Pretentious indeed, for a visitor to the site cannot help but be struck by the untidy, almost riotous look of the place. Scattered about twenty acres of dense woods, with no apparent design, are a group of worn and peeling wood frame bunkhouses and labs, the constant despair of Matt Cornish, the station manager charged with maintaining the place. The visitor should know that looks here are deceiving, for some of the most fascinating and important work in wildlife science has been done at this most unimposing of sites.

I often make a point of showing up unannounced for lunch in summer (the food is excellent). Lunch at the station is always an education. Shortly after the gong of the dinner bell, famished researchers tramp in from the nearby bush and swamps alone or in groups of two or three, stripping off bug nets, hip waders and the thick plaid shirts that are the *de rigueur* dress of the station. The group forms little knots around the tables, debating the arcane points of their science, and I am struck year after year by how little the character of the scene changes. The majority of the crowd are young, fiercely dedicated graduate students, often of the eccentric, "nerdy" type drawn to science. This corps of eager transients is balanced and tempered in both appearance and debate by their greying seers, the distinguished professors of various disciplines of biology who are permanent fixtures of the station. I say permanent almost literally because the crusty wooden walls of the cookhouse are dotted with group photos that date back to the 1940s and track the lives and times of these people in a striking, even poignant tableau. Arrayed in time across this fading gallery are the character-laden images of many of Canada's leading wildlife scientists. The faces one sees most often are those of a close and dedicated fraternity of practitioners of what must be the strangest of biological (or any) sciences – Parasitology.

Parasitology is a weird, complex and indelicate subject. The study, literally, of animals that live in other animals takes one into some of nature's most repulsive and bizarre schemes for survival. Take, for example, *Loa loa*, the African eye worm, a 5-cm nematode or roundworm that commonly afflicts people in the Congo and West Africa. The wretched worm wanders extensively under an afflicted person's skin creating sinuous lumpy tracks as in a sci-fi movie effect known formally as "Calibar swellings" (from Calibar in West Africa where the disease is common). The leech-like creatures often reach the eye in their peregrinations and can actually be seen swimming across a victim's eyeball.[2]

There is something fundamentally obscene about animals that, unlike an honest and decent predator, kill by living within another. People attracted to this odd and unglamorous science do not enjoy the celebrity of more conventional biologists, but unravelling the mysterious and complex lives of these strange animals can be absolutely compelling, addictive to the sort of obsessive personalities that make good scientists.

Roy Anderson, a professor at the University of Guelph and one of the station's icons, was one of these. Professor Anderson (who passed away very recently) was an exceptionally polite gentleman, formal but with an amiable, endearing manner. He engaged generations of students in this unlikely study by regaling them with the life stories of

*A very young moose calf and researcher Rob McLaughlin in Algonquin Park in the late 1970s.* Algonquin Park Museum Archives, Algonquin Park Wildlife Station Staff.

these bizarre creatures with a wit and style of lecture that could at times be pure entertainment. Anderson was professor *emeritus* (meaning that he was too valuable to let go) at Guelph and built his reputation on a piece of scientific detective work that has become known as one of the great feats of wildlife science. He discovered what was killing large numbers of moose in Algonquin.

The moose, along with the wolf and loon, is a symbol of Algonquin Park, and indeed of all the north country. Moose are huge, imposing animals that can appear ungainly but actually move quite gracefully through the thick undergrowth. Born in mid-May, moose calves weigh about 30 pounds but grow rapidly to reach an adult size of up to 1,400 pounds by 4 to 5 years of age. Moose breed, or rut, in early fall and are prominent on the Algonquin landscape at that time, and also in spring when they come out to roadsides to drink the salty meltwater (salt, often in problematic surplus in our diet, is something wildlife often have a desperate need for). By virtue of their size and dark colour, moose are easily visible from the air in winter and are one of the very few species that biologists can actually count or census accurately. We do these surveys more or less annually and the moose population in Algonquin Park has varied from about 2,500 to 3,500 animals for the past twenty years or so.

As early as 1930, biologists from Minnesota to Maine and Nova Scotia began recording observations of moose with a strange and disturbing neurological disorder.[3] Throughout the north country, including Algonquin Park, moose were observed staggering out of the woods in a sort of stupor. The afflicted animals were oddly tame and exhibited a characteristic set of movement disorders. Apparently blind, the moose would stagger around and crash into things in a pathetic reel as with a man drunk. In later stages, the hindquarters would become paralyzed and the head twisted grotesquely to one side. A researcher in Nova Scotia actually witnessed and recorded the slow death of one afflicted moose,[4] but the victims were not often seen to die from the mysterious disease (although this obviously was the end result). Sightings of moose in this state were common enough that the condition became known simply as "moose sickness" and became part of the bush lore of the North.

For years, people wondered about the cause of the mysterious illness and attributed it, speculatively, to dietary deficiencies, bacteria, pesticides and infestations of ticks (more on these later), each theory being closely held by its proponent but all unfounded in science. A virus was also implicated (but viruses tend to get blamed for any disease we don't understand).

Anderson had been intrigued by the problem for some time and suspected a parasite. He had read about a disease discovered by Japanese scientists during the war that caused many of the same symptoms of paralysis and "ataxia" (loss of coordination in walking) in sheep, goats and horses. In the Japanese case, the disease was caused by a roundworm that normally exists in the gut of cattle. The worm causes the cattle no harm but runs amok in the other ungulates (an ungulate is any hoofed herbivore), wandering around and wrecking circuits in the central nervous system, circuits that control movement.[5] Anderson knew of a similar worm in white-tailed deer. Could there be a parallel here with the Japanese case? Was it possible the worm was getting into moose instead of deer, its normal host, and destroying the magnificent animals?

*Parelaphostrongylus tenuis* (whew!) is a pale, slender, 8-cm long innocuous-looking worm – an unlikely candidate to fell a 1,400-pound king of the forest. Anderson knew, however, that as part of the worm's life cycle, it travelled through the spinal cord of deer and into the surface of the frontal lobes of the brain where the worms bred. In doing so, the "brainworm" apparently caused no harm to deer but, based on the Japanese work, it just might in moose.

Intrigued, Anderson experimentally infected two captive moose with brainworm larvae at the research station in the summer of 1963. Within two weeks of infection the moose began to show the first signs of disease – a listlessness and lack of coordination. By three weeks, the legs began to wobble and then to bend grotesquely. The animals developed a crab-like gait, and their heads developed the strange lateral twist. From there, the disease progressed to its endpoint, near total paralysis, and the animals had to be euthanized.[6] Roy had his grim proof, or at least a strong case. It remained, though, to establish for certain that the diabolical little worms were actually causing the disease. Back at Guelph, Roy and his assistants dissected the entire central nervous system and, bit by bit, sectioned the nerves, spinal cord and brain, in search of the worms or tracks of their passage. Like prospectors working a claim, they resisted going right to the most likely site of evidence – the mother lode, the brain – but rather started from the posterior and worked forward, taking sections of tissue every 3 mm. The devastating effect of the worms was soon apparent as the researchers found extensive damage in every region of the spinal cord. The search reached its denouement when the worms themselves were found right in the cranium, many still squirming about in the roots of nerves. Anderson now had undeniable evidence that brainworm was killing moose.[7]

Why the worms kill moose (apparently without exception) but almost never harm deer, their normal host, has never been established but may have something to do with the immune response. Animals have the capacity to recognize the chemical "fingerprint" of invaders (including parasites like the brainworm) and mount an attack usually of aggressive killer-T lymphocytes (a type of white blood cell). When, however, the nervous system is involved, the cure can be worse than the disease. Inflammation caused by the attacking T-cells causes extensive damage to delicate nervous tissues, damage that, in itself, may produce the worst clinical effects. (A very similar and equally misplaced response is what makes meningitis such a critical medical emergency in people.) Moose appear to suffer a more severe reaction to brainworm than deer.[8] (There is also evidence that the worms, "behave" differently in deer and moose, apparently curling, twisting and "digging in" more in moose – this may be a response to the allergic reaction.)

Roy now had his case, but there remained the task of determining how the brainworm got from deer to moose and exactly how the parasite found its way around inside the huge animals. Anderson and his colleagues, including Professor Murray Lankester at Lakehead University in Thunder Bay, Ontario, deciphered the amazing life cycle of the roundworm through a series of trials, observations and experiments in the late sixties.

The journey of the parasite from deer to unfortunate moose is an extraordinary one. The brainworm begins life in the folds along the forefront of the brain of deer. The adults settle in this refuge, "mate" (if such an ardent term can be applied to parasitic worms) and shed eggs into the bloodstream. The eggs are conveyed randomly through the circulation eventually reaching the heart and shunted via the pulmonary artery into the lungs where they develop into tiny larvae. Each larva, now under its own power, migrates up the windpipe to the throat and is swallowed, surviving intact the transit of the deer's digestive system and passing out of its body in the droppings. A snail or slug of any of several species that are common on the forest floor passes by and happens (after their fashion) to feed on the feces and ingest the larva. The larva, awakened from its snug refuge, penetrates into the "foot" or main muscle of the snail and there it rests, awaiting the incredibly unlikely event of being eaten with the snail by a deer (or in this case, unfortunate moose) browsing randomly on vegetation. Released into the gut upon digestion of the snail, the larva, now in its infective stage, penetrates through the stomach of the moose and, in an amazing feat of navigation, travels through the body cavity up into the spine. The rest you know: the infective larva travels the length of the spinal column into the brain developing into an "adult" and wrecking the moose's vital circuitry along the way.[9]

How could such a thing have evolved? The deer, you see, must die and so the parasite must find a way to move from deer to deer before this happens. But how, possibly, could such a primitive organism have become the agent of such an elegant and intricate scheme? Evolution proceeds in imperceptibly slow increments and the life cycle of *P. tenuis* must have developed primitively from a simpler system in a series of reinforcing steps. Perhaps the ancient ancestor of the brainworm was simply a parasite of snails, passing from snail to snail as each died and was consumed by its kin. Some snails and their hidden larvae were inevitably eaten by deer (or their primeval counterparts) and by a fortuitous mutation developed the ability to survive in the deer's gut, benefitting (or, to be more exact, its genes benefitting) from being moved around and broadcast, perhaps, into more favourable environments. In this manner it is possible to dream up schemes for each step on the path to the development of this amazing survival "strategy." Of course, we'll probably never know for sure exactly how the behaviour came about because, unfortunately, behaviour doesn't leave fossils.

In truth, complex life cycles such as this are not at all uncommon in parasites. The odds against the success of any one larva in these improbable ventures are incredible, but most parasites beat the odds by producing prodigious numbers of eggs, so that at least one gets through once in a while.

The odds against individual parasites "winning" – that is getting into their target host – are nevertheless so great that some have evolved incredible means to succeed. Parasites which have an "intermediate host" (like the brainworm's snail) often "manipulate" the behaviour of the carrier so that it is more likely to be picked up by the final target host. The best known example of this is a flatworm, or liver fluke, of sheep which uses ants as its carrier. The larva of the flatworm infects the brain of ants, causing the ants to behave erratically and to be more likely to place themselves in the pathway of grazing sheep! (Of course the larvae aren't consciously "manipulating" anything; they are merely performing as their genes dictate.)

Tom Nudds, another researcher from Guelph, and a graduate student named Karen McCoy figured that the brainworm had to employ some similar dodge to get into deer. The prevalence of the worm in deer is so high (relative to its occurrence in snails) that the two researchers reasoned that the worm must be somehow "manipulating" the snail to beat the odds. They studied the climbing behaviour of infected and uninfected snails, anticipating that the infected snails would climb (as the saying goes) higher, faster and farther. They were disappointed; infected snails turned out to be no better climbers than "clean" ones. The brainworm apparently relies on chance alone to infect moose.[10]

My most memorable encounter with the brainworm (or at least its consequences) happened on Redrock Lake in the south-centre of the Park in the late fall of 1997. Fisheries researchers working on Redrock had noticed an adult bull moose lying in the cold shallows just offshore; a very unusual thing to see that time of the year. They returned several days later to find the moose in exactly the same spot and, now thoroughly concerned, they reported the sighting. I made my way to the scene the following day, accompanied by a park ranger.

*A bull moose in its prime, dying from brainworm.* Courtesy of Norm Quinn.

The moose, a huge, near prime bull, was exactly where they had reported, lying on its side near shore in about two feet of water. One sees a lot of nature's cruelty as a biologist, but the sight of this magnificent animal in such a pitiable condition was heart-rending. The moose was moribund, unresponsive and wasted, and weeping thick pus-filled tears from one eye. It had been browsing pathetically on some brush on shore that was just barely within reach. The moose was unable to rise on its hind legs despite our prodding – the classic paralysis of brainworm. We had to shoot the animal and I spent the trip home ruminating on the random brutality of nature.

Moose are majestic animals, roaming the north woods with a haughty disdain of all but wolves and hunters. Yet these magnificent animals fall victim to another diminutive but insidious killer: the winter tick. The winter tick *Dermacentor albipictus* (whew!) is typical of its loathsome kind – eight-legged arthropods that have plagued mankind for millennia as carriers of an array of serious disorders like Lyme disease. Ticks, as we shall see, can have an equally devastating effect on wildlife.

The winter tick of moose has a much simpler life cycle than the brainworm; a "direct" cycle in the jargon, that is, transferred directly from one victim to another with no intermediate host. The tick, normally unseen, reveals itself and its shocking damage in late winter or early spring and thus the name.

Winters are long in Algonquin Park and just about the time that we begin to enjoy the return to warmth, the long, brilliant days of March thaw, we normally also see one of the north's most extraordinary spectacles. Moose appear at roadsides in a ghastly state: emaciated, almost denuded of hair and often too weak to move if the disease is in its terminal state. The hair loss reveals the underlying pale skin, giving the moose a ghost-like appearance (thus the term "ghost moose" is given to the worst cases). The condition results from the infestation of incredible numbers of ticks that become so irritating that the moose, in a torture of itching, lick and rub their own hair off.

The ticks at that point are revolting creatures, each being about the size of a quarter and engorged with blood. They appear absurdly, even from a distance, as grotesque clusters of grapes hanging off the animal's hide. The "adult" (again, the term seems oddly inappropriate) females, engorged with blood, drop off the moose in spring and lay their eggs under rocks and leaves. The eggs hatch and the tiny larvae lie low until the first cool days of fall when they ascend vegetation and "quest" bizarrely with their claws, groping the air in hopes of a moose passing by. Moose normally acquire a modest and tolerable load of these "questers." The larvae, however, are not distributed randomly in the bush and tend to be concentrated in spots where, say, a moose lay down for a period in spring. Moose, for reasons known only to moose, tend to pick the same places to rest, and if an unlucky animal happens to lie in one of these hot spots in fall, he or she can acquire incredible numbers of larva.[11]

Each larva (which look like tiny spiders) rests in the snug underfur and develops into a tick by mid-winter.[12] The aggressive blood feeding of thousands of ticks drives the moose mad and the animals obsessively groom (lick) themselves, scratch with their hind hooves, and rub furiously against trees to relieve the irritation. A severely afflicted moose will actually rub large areas of hide raw and, once this happens, with the loss of insulation the moose enters a spiral of energy loss that may end in death. It appears that the animals simply cannot eat enough to replace the energy lost to the bitter northern cold and, essentially, starve.[13] Before actual starvation, however, it is likely that the symptoms of severe stress will set in as the immune system is defeated and the animals may succumb to aggressive infections like pneumonia.[14] (Incredibly, some ticks actually inject chemicals into their hosts that "switch off" the allergic irritation and itching, thus protecting themselves from the hosts "grooming.")[15]

We routinely do counts of moose from the air in winter, and it was during one of these surveys that I had a remarkable encounter with the winter tick. Two conservation officers and I were near the end of a long day of flying on a bitter cold day in mid-February when one of the officers spotted a cow moose with a calf. We circled the moose, confirming the observation, and turned to move on. Something about the scene, however, had caught our subconscious attention and, without even asking, the pilot turned back. The moose were still there and that was what had alerted us; they (or the calf at least) was *exactly* still there. The animal had not moved an inch and seemed frozen on the spot, almost riveted into the ground. We circled the scene for some time, several times "buzzing" the animal to get it to move. Gradually, however, we came individually to the conclusion that the animal had died standing up and one of the officers eventually and rather sheepishly suggested just that!

Now, conservation officers tend to be extravagant fellows and given to exaggeration but, by the end of the trip home, I had to admit that it seemed the only explanation.

The situation was unusual enough that we resolved to return to the scene the next day. The trip was tough, with long miles forcing snowmobiles through dense bush and deep snow, but we found the calf alone (the mother moose having apparently given up hope and left) and indeed stiff and dead but fallen over. We dragged the moose out with great difficulty and I drove it the next day to our research lab in southern Ontario. The animal was necropsied (the veterinary equivalent of autopsy) and it was determined, by a sampling process that I neither really understood nor wished to take part in, that it had carried some *fifty thousand* ticks. Upon skinning, the true extent of the calf's suffering became apparent, for it was so thin as to appear not much more substantial than a large dog.

Unlike the brainworm, the devastating effects of which are rarely seen, the winter tick can have spectacular effects. There have been two such "die offs" in recent years in the Park. There is a theory that outbreaks

*A young victim of the moose tick.*

*A moose skeleton found near Crowe Lake by David and Tim Young. The skeleton is intact, which is extremely unusual, suggesting that the animal died of disease. Predators scatter the bones widely about the area.*
Courtesy of Tim Young.

of winter tick are controlled by weather in late winter or early spring.[16] When spring weather is harsh, and particularly when snowmelt is late, the female ticks fall onto snow upon leaving the moose. Many die in these circumstances[17] and the production of eggs, and thus larvae the following fall, is poor. The reverse may happen in "good" years with mild and early spring weather. Under these conditions, the survival of egg-bearing females is presumed to be good, there is prodigious production of young "questers," and during the following winter moose die in droves.

In the spring of 1989, as the Park opened and canoeists and fisherman began exploring, we began to receive report after report of dead and rotting moose. These sightings increased in frequency into May and it was soon obvious that something unusual had happened. Ed Addison, a wildlife disease expert with the research section in Maple, came up to investigate.

Ed is something of a rarity: a scientist who is a real "people" person. A great communicator, Ed revels in the opportunity to speak to naturalists, trappers and hunters about his research, being especially fond of giving "tick talks" (giving a measure of his sense of humour). Ed and I spent several days that spring walking trails seeking out reports of moose carcasses. It soon became obvious that a disaster had occurred: dead moose were everywhere, the scent of putrefying flesh creating an almost macabre atmosphere.

On one such sojourn, I waited uncomfortably at a carcass for about an hour as Ed did his investigation. I say uncomfortably because the carcass and its immediate surroundings were crawling with ticks and the revolting creatures did not distinguish us (and most importantly me) from moose. Now, wildlife specialists have a sort of macho image to uphold and, not wanting to sully that tradition (especially in front of Ed, whom I greatly admired), I toughed it out. But every fibre in me cried out to leave, especially as I began gingerly picking ticks off my boots.

We went on in this manner for some time and had checked out five or six carcasses by the end of a long day. Exhausted, we split up and I began the long drive home, anticipating a hot shower. Before long I began to feel a strange pinching sensation on my lower legs and then, shockingly, my nether regions. Horrified, I pulled over, streaked into the bush, disrobed and discovered the little brutes *everywhere* (if you get my meaning). I even encountered one on the shower drain that evening. Having these appalling creatures on you is, trust me, an unsettling experience and a most graphic illustration of their tenacity.

Ed Addison conducted research on the winter tick at the Wildlife Station in the mid-'80s. Much of the work involved long hours watching captive moose from a makeshift observation tower and recording how they tried to cope with the nasty little brutes. These observations were, of course, done in daylight hours. One evening at supper, Addison, ever the perfectionist, announced that, in the interest of good science, the observations would have to continue through the night. Furthermore, he, to set an example, would take the first shift. The research team were taken aback but, being young students and as dedicated as Ed, they were nevertheless keen to comply and a schedule

*Feeding young moose at the Wildlife Research Station during the period of the tick study, Sasajewun Lake, 1980.* Algonquin Park Museum Archives #5689.

was hastily drawn up. Addison took the first shift at sundown. At the appointed and ungodly hour of 4 a.m., his relief arrived at the tower to take his place. The young fellow announced his presence but to no avail; there was no response. Alarmed, the student climbed the tower to find Addison curled up in a corner, sleeping contentedly. The night work was cancelled.

This setback notwithstanding, Addison's work contributed hugely to our understanding of how the winter tick affects moose. It had gradually become clear that, unlike most parasites which coexist with their host, the winter tick can devastate moose populations. There are well-documented cases in western Canada of the tick causing serious and prolonged reductions of moose numbers.[18] A very recent study on Isle Royale, Michigan, showed a relation between tick numbers, a chemical in the blood that measures the physical well-being of moose, and the actual numbers of moose in the population.[19]

Effects on Algonquin's moose, however, have been less clear. The Park's moose population actually increased slightly after the '89 event. In the mid-'90s, a researcher did an analysis of trends of Algonquin moose numbers and found no connection with the severity of tick infestation[20] (which is periodically measured by an aerial survey in late March). More recently, though, effects have been dramatic. Consecutive die-offs in the spring of 1999 and 2000 were devastating, reducing the Park population by half.

In all likelihood, the Algonquin moose population will recover in short order. As Canadians, we know perhaps better than anyone that weather can be fickle, positively perverse at times. However, irregularities of weather and the tick-friendly warm springs we've seen lately are considered random events, episodes that are not likely to continue. We can look forward to the return of chilly Aprils, and our moose, very soon. Having said that, one has to be concerned (if one is a moose) about the apparent warming trend in global climate. Warm springs may be the harbinger of a bleak future for Algonquin's moose. Moose, in fact, do not enjoy warm weather at any time, suffering grievously on hot summer days. However, as with much in wildlife science (and climate warming itself, for that matter), the link between spring weather and moose die-offs is a tenuous one – an appealing but unproved hypothesis at this point.

The brainworm too is a scientific enigma. It obviously kills moose, but ever since Roy Anderson's groundbreaking work on captive moose there has been an active debate about whether it really kills that many moose in the in the wild (partly because, as discussed earlier, moose infected with brainworm are so rarely seen). There are places (New Brunswick, for example) where deer and moose coexist with both at fairly high numbers; given what we know of the malignance of the brainworm to moose, this should not be possible. Attempts to resolve this paradox (how moose can live with deer in the wild) have been thwarted. We know how the worm functions from individual deer to moose, but we don't understand its effects in nature where thousands of deer and moose, billions of snails, and their respective cycles, habits and habitats make things so complicated.[21]

These sorts of problems illustrate the difficulty in looking for hard and fast rules about wildlife and their mysterious cycles. Anderson, Addison and all the others can, despite years of focus on one question, only scratch at the surface. Nature, in its vast complexity, has a way of thwarting our attempts to look for broadcast solutions. Parasites, when you think about it, should not kill their hosts; they need them to survive. As a rule this is true; most parasites live in harmony (if often a discomforting one) with their host. The winter tick and brainworm (in the wrong place) do not, often with tragic and spectacular results.

*A young bull moose in its prime. This one has escaped ticks and brainworm.*
Courtesy of Peter Smith.

Wildlife parasites are not, as a rule, transmissible to humans and the meat of deer, moose and other popular game animals is generally safe to eat. A notable exception is yet another parasite of moose but whose main effect, in this case, is felt by wolves. This is the tapeworm known to science as *Echinococcus granulosus*. Tapeworms are, even for parasites, repulsive creatures, but *E. granulosus*, is one whose potential effect is as gruesome as its appearance.

The worms exist as adults in the digestive tracts of wolves where they "mate" and pass eggs in the wolf's feces. The eggs hatch upon being consumed inadvertently by a grazing moose (again, consider the odds!), enter the circulatory system as larvae, and normally come to rest in the lungs where they develop, over time, into enormous cysts.[22] People who live around wolves, and particularly wolf researchers, are warned (unnecessarily one would hope) not to sniff or consume (!) wolf droppings because humans can act, in this case, as perfect surrogates for moose. Normally, the growths (known as "hydatid cysts" and thus hydatid disease) are commonly found in the lungs, but they can develop almost anywhere (in bones for example, which they can weaken and cause to fracture) and must be removed surgically. Therein, however, lies the biggest threat, for the surgeon must be meticulously careful not to puncture the cyst in removal. Escape of any internal fluid can cause anaphylactic shock[23] (the severe allergic shock more commonly known with things like peanut allergy). Perhaps worse, particular bits of the cyst may, with all but the most scrupulous surgical care, be broken off and shed into the bloodstream where they will spread like weeds, developing into dozens of new cysts anywhere including, tragically, the brain. Don't sniff wolf dung.

We are often asked to "do something" about these things. There are veterinary options – drugs, for example, that can eliminate almost all parasites with one dosage. But this is the real, and at times, very cruel natural world. Parasites can achieve fantastic numbers and are almost everywhere in the wild. Eliminating things like the winter tick is, in principle, possible but the ecological effects of the widespread use of pesticides that would be necessary would be devastating. Besides, parasites and their works are part of nature's scheme and, in Parks at the very least, should be allowed to function unimpeded. As loathsome as they are, they are part of the system.

# 3 | Hemlock and History

*No living man will see again the virgin pineries of the Lake States, or the flatwoods of the coastal plain, or the giant hardwoods.*[1]
— Aldo Leopold, 1966.

Go back 400 years. North America was a paradise – a vast and varied tableau of virgin forests, grasslands, tundra and desert. Each of these pristine ecosystems was sparsely occupied by one of dozens of Native cultures adapted through thousands of years to survive in its unique offering of opportunities and hazards. European man had secured only tenuous toeholds in Florida, New England and Quebec, and had yet to impose his unique management style on the land. The variety of this utopia was striking, ranging from steamy Everglades and dense rain forests to vast grasslands and barren, frigid tundra. North America's great Parks were established, beginning with Yellowstone in the U.S. in 1872 and in Canada with Banff in 1887 (and Algonquin in 1893) to protect precious remnants of these ecosystems as windows or "vignettes" into the primitive past.

One of the more prominent of these ancient landscapes was an immense expanse of majestic forest that stretched from Wisconsin to Maryland and Maine to Missouri. Algonquin was at its northern tip. This deep and shadowy forest was seemingly unbroken and without limit; it is said that a squirrel could travel from treetop to treetop from Detroit to New York and never touch the ground. This was the great barrier faced by the pioneers, an impenetrable wall known today technically as the Northern Hardwood Forest.

It is easy to come to think of the rolling farmlands of places like upstate New York and southern Ontario as "natural"; the cornfields, dairy farms and woodlots are so familiar and seem so integral to the land as to be the permanent and proper condition. In truth, were nature ever to have its way, the pastoral scene would give way in time to a completely different setting. The open, spacious landscape of rolling hills, rich farmland and scarlet sunsets, so comforting to the eye in its own way, would be replaced by a great expanse of dense, dark and foreboding cathedral-tall forests.

The original forests of the Northeast must indeed have seemed ominous, even evil places. Few authentic accounts survive but one early Ontario pioneer described a forest: "through which the sun never penetrated."[2] A settler in Beckwith township, Lanark County, in eastern Ontario left the following powerful description: "There is something in the ponderous stillness of these forests, something in their wild, torn mossy darkness, their utter solitude and mournful silence which impresses the traveller with a new aspect each time he sees them .... In Upper Canada the endless hills of pine give way at last, or at most stand thinly intermingled with gigantic beeches, tall hemlocks and ash, with maples, birch and wild sycamore, the underwood of these great leafy hills. Mile after mile, and hour after hour of such a route was passed – a dark black solitude, with here and there a vista opening up, showing the massive trunks, grey as cathedral ruins, which bore the rich canopy of leaves aloft."[3] Still another early immigrant described the forest with obvious apprehension as "moors, for instance, all grown over with large trees, some fresh and green, others half rotten and a great many rotten from top to bottom and almost as many lying in all directions as are standing, with not a living creature to be seen or heard except a bird or two, and the owl screaming in your ears at night."[4] Finally, the great Henry David Thoreau, who was if nothing else a keen observer of nature, left the following description of how the younger forest had changed from its primitive state: "It has lost its wild, damp and shaggy look" and, "The countless fallen and decaying trees are gone, and consequently that thick coat of moss which lived on them is gone too. The earth is comparatively bare and smooth and dry."[5]

The forest, you see, was mostly "old growth" and as much dead as alive. Forest fires, the main agent that keeps woods young, had a hard time getting started here. The humid weather of the Northeast and dense, fire-resistant trunks of the dominant hardwood trees prevented fire from establishing and "cleansing" the forest. Trees simply grew up, reached a very old age and died, littering the forest floor with their corpses and creating infrequent gaps for aspiring followers in the understory to exploit. The woods were thus a dense and complex concert of young and old and dead and dying trees atop a tangled mass of rotting debris. This, beneath the cathedral-like canopy of centuries old Hemlock, Maple and Beech would have created a dank, shadowy, awe-inspiring scene. Places like (or very near to) this can still be found in parts of Algonquin Park.

But this is not even close to the whole picture for, on a broader scale, the scene was vastly more complex. Our squirrel, in fact, could never have made his trip without a few detours for nowhere, really, were the woods completely intact. The forest, being old, decadent and weak, was not windfirm. Frequent random and violent thunderstorms and occasional tornadoes would flatten the woods, creating various-sized patches of new growth.[6] These new forests were normally of a different type entirely, being composed of sun-loving poplars, birches, spruce and sometimes pines. On upland sites, which were better drained and drier, stands of pine took hold. Pines (and coniferous trees in general) burn easily and lightning fires frequently burned through the majestic white pines, creating more patches of young growth scattered erratically across the landscape.

The Native Peoples were also key players, for some tribes purposefully burned the forest to clear the land for planting and as a form of game management. In places (but probably not in Algonquin Park[7]) burning by Indians was so common it is thought to have transformed entire landscapes.[8] (Actually, the amount of burning that Native Peoples did is "hotly" debated by anthropologists.)

We see, then, that the eastern forests, although ostensibly a simple mantle of green, were in fact highly complex mosaics. In the mid-1930s,

*On June 15, 1974, Paul Keddy, a former summer naturalist at Algonquin Park, discovered a new plant, ultimately named* Dryopteris x algonqinensis, *by the shore of Greenleaf. This plant was the first and only plant of its taxon for the world. The plants are believed to still be there. Left to right (rear): Dan Strickland, then the Visitor Services Planner for the Algonquin Region, pointing towards the plant; Tony Reznicek, formerly of the MNR, and then a doctoral candidate in botany at Erindale College; front: Paul Keddy; Dan Brunton, then the Environmental Planner for the Algonquin Region; and Dr. Don Britton, botanist from the University of Guelph.*
Algonquin Park Museum Archives #2120, MNR.

a handful of farsighted botanists and foresters, recognizing what had been lost, became determined to document for posterity the pathetically few patches of virgin forest that remained before their ultimate demise. Botanists entered into these precious remnants (the best of which were in, of all places, Pennsylvania) and described the flowering plants, herbs, shrubs and trees in exquisite detail from the ground up. The extraordinary diversity of this lost world was soon evident, prompting Sam A. Graham, a leading forester of the time, to state laconically that "it seemed evident that the forest varied greatly in composition from place to place."[9] G.E. Nichols of Yale, another perceptive early conservationist, captured the fundamental majesty of the scene with "one may picture a forest in which broad-leaved trees formed a dense stand from eighty to one hundred feet high above which, either by small groups or single trees White Pine reached to 150 feet or more."[10] We know from the work of these early researchers just how beautiful and varied the forest was.

The original articles, in which these inspired efforts are written, can be found with due patience in the dusty depths of the library of any major university. One is struck, in perusing the old papers, by the elegant and lyrical names botanists assigned to the plants they discovered and indeed the often charming names of the botanists themselves. Plants with enchanting and vaguely erotic names like "Maidenhair Spleenwort," "Golden Corydalis" and "Hairy Bearded Tongue" were named by the likes of "Linda Llewellyn" and "Wendell Weatherwax." (It is odd, is it not, how a person's name often reflects their work; I know an ornithologist named David Bird, a wildlife biologist called Roger Wolf and a fisheries biologist by the name of Peter Perych.)

Another way that botanists have reconstructed what the original forests looked like is through the pollen record. Pollen, which falls in prodigious quantities from trees every year, often, of course, falls on lakes where it sinks and accumulates on the bottom. On a few very deep lakes the bottom is never disturbed and the pollen forms bands over time, much like the growth rings of trees. These strata can be "mined" by coring and experts can then read the history of pollen and entire forests through eons of time. There are two of these "meromictic" (never mixing) lakes in Algonquin Park and the record has been read on both.[11]

As complex as the pristine forests was the community of wildlife that inhabited them. Dozens of species of birds alone had evolved with these forests, shifting in like-minded assemblages to exploit new habitats as they developed. The Northern Hardwood Forest was one of the most complex of the spectacular pristine landscapes that comprised the New World.

All this came to an end in an astonishingly short time. From about 1750 to 1930 the forests of eastern North America were, for all practical purposes, *completely* destroyed. Loggers assaulted the woods, initially to clear land for farming, but later to harvest White Pine almost exclusively

*John Rudolphus Booth of Ottawa was Canada's foremost lumber baron. In a quote from* The Lumberjacks, *a description attributed to "the* American Lumberman *in 1904 ... estimated that Booth's 4,250 square miles of timber limits on the Ottawa watershed were 'sufficient timber land to make a strip a mile wide reaching across Canada from the Atlantic to the Pacific.'"*[12] Ontario Archives, Publisher's Collection.

*Loggers were housed in shanties about 40 feet square, constructed of logs dovetailed together and chinked with moss and mud.* Ministry of Natural Resources, Publisher's Collection.

for the massive timbers needed for shipbuilding. G.G. Whitney of Harvard University points out succinctly that "the full weight of a technologically advanced culture was thrown against the pine forests of the Interior of the continent," and quotes the cutting invective of conservationist Arthur Lowers that "the sack of the largest and wealthiest of medieval cities would have been but a bagatelle compared with the sack of the North American forest."[13] The great climax Pine-Hemlock forests of the Great Lakes region were reduced to a few pitiful fragments in less than 200 years.

Algonquin did not escape the onslaught. As early as the 1830s, loggers entered the Park, most probably from the east along the Petawawa and Bonnechere rivers which provided ready access via the Ottawa River to Quebec City, in quest of the Park's magnificent White Pine. In spite of appalling working conditions and the fearful physical challenges presented by the wilderness, the despoliation of the Park's virgin forests progressed briskly; by the 1870s upwards of 70,000 sawlogs and 15,000 pieces of square timber were floated out on river logdrives each year. The object of all this effort was Pine. The Park's stately White and Red Pine, perfect for the making of massive timbers, were felled, "squared" on site to fit in the holds of ships, floated downriver

*A rare photograph of a typical camboose shanty, which would have been found in the area that became Algonquin Park, circa 1850s. By this time some thousands of men were living in these shanties within the many timber limits.* Algonquin Park Museum Archives #6811 – Henry Taylor.

*Those who worked in the woods all favoured hiring on with camps whose cooks had good reputations. The interior of this cookery shows the cook and cookies employed in a Whitney camp at the eastern limits of the Park.*
Ontario Archives, Publisher's Collection.

and shipped to England[14] (where, ironically, they were made into more ships).

The men thus employed faced brutal living and working conditions – conditions that today would spark a judicial inquest. Leaving their farms in fall, the men trekked for miles into the Park to live and work for months on end in bitter cold and dreadful living conditions. They were housed in gangs of 50 or more in cramped, gloomy, lice-infested log shanties called "camboose camps," neither washing nor changing clothes for weeks.[15] The loggers lived almost exclusively on a diet of brown beans (the consequences of this and unwashed bodies on air quality of the shanties being unthinkable). Living in such squalor and working with crude tools in deep snow, cruel cold and constant hazard, it is hard to believe that these men could have levelled Algonquin's pine forests in a few short decades. But that they did, for not long after the turn of the century most of the Park's virgin pine woods were gone.

There followed a period in which the focus turned to hardwoods as timber markets ebbed and flowed. The famous Mosquito fighter-

*Henry Taylor (left) and his older brother Jim of the Madawaska Valley area had an association with Algonquin Park dating back to the lumbering camp era. In their later years, they successfully set out to locate reminders of Algonquin's lumbering heritage. Their finds included this old decaying wooden dam.* Ontario Archives, Publisher's Collection.

*Henry Taylor, a noted craftsman, built canoes, chairs and beautifully woven baskets, as well as carving primitive art pieces, many of which are housed in museums. Bancroft named Henry Taylor "Citizen of the Century" in 1999.*
Publisher's Collection.

bombers of World War II were, for example, constructed entirely of yellow birch plywood from Algonquin Park.

Then, in 1962, someone in Toronto decided to build a subway. The import of this decision on the environment and wildlife of Algonquin was, as we shall see, momentous. Hemlock, unfortunately, will not rot underground and makes excellent shoring timbers for subway construction (it is also used, oddly enough, as a medium to grow mushrooms, a foodstuff that, once scorned, has become a trendy favourite of the yuppie set). Through the 1960s the focus of loggers turned to hemlock and much of the hemlock (particularly that in important deer wintering areas) in the Park was removed.[16]

All of this application of man's industrial prowess had a huge impact on the Park's wildlife (as logging did throughout the Northeast) but the effect on the Park's ungulates (moose and deer) was profound. The changes wrought by logging determined the ebb and flow of Algonquin's moose and deer, but the two have had a form of revenge for, as we shall see, they in turn engineer the forest.

But what do we really know of the history of moose, white-tailed deer and others of their kin in the Park?

Knowledge of the distribution of big game animals in central Canada prior to the arrival of the white man is surprisingly scant – in fact, almost absent. One reason for this is that there is little evidence from archaeology to draw on; bones rot rapidly in the acid soils of the Canadian Shield and were only occasionally preserved if charred in cooking fires by the Native Peoples. Nearly all we know about the original wildlife community is based on educated guesswork and the foggy writings of early explorers.

Jacques Cartier, who explored the St. Lawrence Valley in the mid-1500s, is perhaps the first reliable narrator. Cartier wrote about seeing "*cerfz*"[17] (a sort of generic term in French for "deer" that could have meant deer, moose caribou or elk). Cartier also observed the Natives hunting "daims," a term that some have interpreted to mean moose. There is an account from 1606 in New England, from an excursion up the Sagadahoc River of an animal larger than deer called "*Mus*"[18] that was, presumably, a moose. There is also an early reference to "caribou" in the vicinity of North Bay, north of the Park. The early accounts, however, are extremely confusing; a Jesuit priest who lived with the Ottawa and Mohawk Indians in the 1600s left an artful sketch of a stylized animal that is clearly a moose but labelled a caribou.[19] Champlain wrote of Natives hunting "elk" and later "*orignacs*," the latter term being a precursor of the French word for moose. Various maps of eastern Canada exist from the early explorations with the words "cerfz" or "stag" scrawled here and there. These early accounts could, however, mean almost anything. The term "elk" is a good example of the confusion. "Elk" is used by Europeans to describe the moose that inhabits Scandinavia. Consequently, the early Europeans called our moose by the name elk.[20] The word elk, however, eventually came to apply to the majestic animal best known from the Rocky Mountain foothills that should really be called "wapiti." "*cerfz*" and "stag" in the early accounts could really have meant any male specimen of deer, moose, elk or caribou. Nevertheless, it does seem almost certain that the early observers saw moose in New France, the climate and forest habitat of most of which is similar to Algonquin Park.

The early observers notwithstanding, hard evidence is almost totally lacking. Excavation of a Native campsite at Whitson Lake on the Park's east side in 1972 turned up, intriguingly, a number of charred deer bones.[21] The site, however, could not be dated and the author speculated that the campfires were not very old, and certainly postdated the initial European invasion. An analysis (or more accurately, educated guesswork) at the time came to the conclusion that deer were, in fact, probably absent from the primeval Algonquin scene.[22]

It is quite possible that caribou were on the early Algonquin scene, but there is a problem. Woodland caribou require coniferous forests that support extensive growth of tree-borne lichens. (Lichens are the colourful moss-like plants commonly seen on rocks and trees – they are actually a combination of fungus and algae.) The forests of Algonquin do not support the crucial lichens in sufficient abundance and probably did not in the past (incidentally, how caribou survive for months exclusively on lichens, which have almost no nutritional value, is a complete mystery). That said, the late Ralph Bice, the venerable trapper and celebrated seer of Algonquin Park, relates intriguing stories he heard from

*Ralph Bice, the "grand old man of Algonquin," is shown here at age 93, joining in the ribbon-cutting ceremony with Bob Rae, premier of Ontario at the time. The occasion was Algonquin Park's Centennial Celebration, 1993.* Ministry of Natural Resources, Publisher's Collection.

his father and others of caribou antlers found near the Park in the early 1800s,[23] and a caribou toe bone was found in a dig in Middlesex County in southwestern Ontario in 1939 (its identification, however, is disputed).[24] We know that caribou originally ranged through much of northern Ontario and they are also known to have existed prehistorically in Maine. So caribou (and elk too for that matter) were possible players on the early scene. But we don't know for sure.

Moose are a certainty. Algonquin is prime moose country and there is no reason to think it was not so at the time Columbus landed. The advancing edge of the last great glaciation forced moose south to take refuge in what is now New England and the general area of Missouri. As the glaciers receded, moose returned to exploit the new forests developing in their wake,[25] and moose almost certainly reoccupied Algonquin Park several thousand years ago.

Deer probably found Algonquin, and much of central Canada and the northern contiguous states, a bit hard to handle. There are two types of deer in North America: the mule deer of the rugged western foothills and the familiar white-tailed deer of our eastern forests. The white-tailed deer is an extraordinarily successful animal. Although it prefers forest habitats, the white-tail thrives in environments as fundamentally different as the Florida Everglades and Saskatchewan prairie. The white-tailed deer has, in fact, adapted and indeed flourished with man's incursions and is likely more abundant now than prior to settlement. The harsh winters of Algonquin Park, however, probably presented an intractable barrier to deer.

Winters in Algonquin are long and hard. Snow generally forms a permanent cover in late November and lasts well into April. Temperatures dip into the minus twenties or even thirties for weeks on end. Deer, unlike moose, are not well-equipped to deal with such severe conditions. Having shorter legs and a much slighter and less energy-efficient build than their larger cousins (moose, elk and caribou), deer struggle in deep snow and cold. Their efforts to find the woody twigs of shrubs and trees, which is the only food available to them in winter, can be completely draining. Deer can actually use more energy in desperate search of food than gained from the food and, faced with this untenable situation, enter a slow spiral to starvation. Hard winters can decimate deer populations; it is not unknown for deer to suffer losses of two-thirds over a severe winter (something, actually, we have not seen in the Park for some time). The Park can at times be a pitiless place for deer.

There is, in fact, very little evidence to support the existence of deer prehistorically north of the Great Lakes. The great naturalist Ernest Thompson Seton did not mention deer in his estimations of the early fauna of Manitoba[26] and southern Manitoba suffers winters similar to those in Algonquin. Deer bones were found in excavations of Native camps in Huronia, southwest of the Park,[27] and also south of Ottawa,[28] but the dates are uncertain and the weather in those parts of Ontario is significantly milder than in Algonquin. Remains of deer have also been found in Indian middens (refuse heaps) in Maine, prior to the disappearance of caribou in that region.[29] But, again, weather in Maine is less extreme. The balance of evidence, and best guesswork, is that deer were not around the early Algonquin scene.

This changed dramatically in the late 1800s. As loggers cleared the bush, extensive areas of regeneration or "second growth" grew up,

*Deer were abundant in Algonquin Park through the early to mid-1900s, including the early 1930s when this picture was taken by John W. Millar, the Park Superintendent from 1924 to 1930. This photo of startled deer was taken near Cache Lake.* Algonquin Park Museum Archives – John W. Millar.

providing the thickets of woody browse that deer need for winter food. This, adjacent to the original havens of hemlock (the conifer that best shelters deer in winter) provided the right combination of food and cover, and deer appeared for probably the first time. There are reliable records of deer in the area that ultimately became the Park as early as 1860[30] and the deer population grew steadily into the new century, an early estimate putting the Park's population in the tens of thousands. There is thought to have been a "crash" in the late twenties but, with the accelerated opening of the forest, the population grew rapidly afterwards and, by the late fifties, deer were plentiful.[31] R.C. Anderson, of brainworm fame, told the author of having "hundreds" of road-killed deer to dissect for his research in the early '60s. By then, deer were extraordinarily abundant and commonly seen by the dozens along Highway 60, the Park's busy southern thoroughfare. It is thought that there were as many as twelve deer per square mile in the Park in the late '50s.[32]

Not surprisingly, given what we know about the lethality of the brainworm, moose were relatively scarce in the Park during the heyday of deer in mid-century (as a measure of this, consider that during a beaver survey in 1939-40, a researcher counted 254 deer but only 19 moose).[33] There are in fact few records of moose through the revival of deer and, although there are skeptics, Professor Anderson was convinced that it is no coincidence: the invasion of deer brought with it levels of brainworm that moose could not tolerate.

Around the early 1970s the situation began to reverse. Logging in the Park became less intensive and the extensive pastures of browse grew out of reach of deer (moose, of course, need this winter food also but are more mobile than deer in snow and can access it farther from coniferous cover). Coincident with this, construction of the Toronto subway thinned out many of the Park's hemlock stands. There followed a series of severe winters and deer suffered dreadfully. By the early 1980s, deer had all but disappeared from the Park and moose (surprise, surprise) became extraordinarily abundant.[34] From nearly undetectable numbers in the '60s, the moose population grew to about 1,500 in the mid-'70s and to about 3,500 by 1980 where it has remained, more or less, since. (All this, incidentally, had a significant impact on the Park's wolves but that story is for later.)

For an impact of logging on wildlife of an entirely different sort, consider the following story that I have heard so many times around the Park that it must be true. It seems that one winter in around 1948, just after adoption of the chainsaw into common use, a logger was applying this new toy to a massive white pine. Halfway through the trunk he, the tree and surrounding snow were suddenly drenched in blood coming off the saw blade! Shocked, and assuming he had cut himself, it took the man some time to realize that he had sawn a hibernating bear nearly in half! Large old pine trees can often develop a hollow rotten core from an earlier wound or infection, perfect spots for bears to den in. The hapless logger had approached the tree from the sound side and was not made aware of the bear's presence until it was too late. The bear did not survive but being in a state of torpor did not, presumably, suffer excessively!

We see then that moose and deer numbers in the Park have fluctuated with the state of the forest and weather. The events I have related are the commonly accepted history and seem, superficially at least, fairly straightforward. The details, however, can be considerably more complicated as most questions around wildlife and nature are.

Hemlock may be the key! The eastern hemlock is not as striking a tree as the grand and lofty white pine of greater fame, but it is of huge importance in the forest and particularly to wildlife. This reluctant supporter of both subways and mushrooms is, in fact, the single most important wildlife cover tree in the northern hardwood forest. The hemlock's dense crown of flat needles holds snow better than any other coniferous tree, and much of this snow melts (or more properly "sublimates") directly into the air on warm sunny days. Snow, therefore, does not accumulate as deeply under hemlock trees; the ground beneath can be nearly bare even in mid-winter, providing a vital refuge for deer and other wildlife. Hemlock is a preferred cover tree for an impressive array of wildlife, some 30 species of birds alone, and the presence of even a single hemlock in a hardwood stand will add notably to the variety of wildlife.[35]

In the early 1980s, foresters in and around Algonquin Park gradually came to a disturbing realization – hemlock was disappearing from the forest. Forestry surveys and the common wisdom of woodsmen showed that there was virtually no young hemlock "recruiting" or growing up from the forest floor, even where mature hemlock trees were nearby to cast seeds. Hemlock is difficult to grow, it requires very particular soil and moisture conditions, but the apparent total failure of reproduction was a mystery. The tree was, after all, known to be common in the original forest so must have the basic means for success. Why, then, was it failing now? Hemlock is incredibly long-lived, surviving up to 500 years, so the remnant trees, at least, will be around for some time. But the nearly complete failure to reproduce was an ecological dilemma and, for the future forest, a calamity. What was happening? The enigma had both intrigued and frustrated Park foresters for years.

And then along came Stanley Vasiliauskas. Stan, a botanist, arrived on the Algonquin scene in 1992, taking on the hemlock question as the subject of his Ph.D. thesis. Stan is a highly intelligent fellow with piercing dark eyes and an ever-present smile. Affable, but reticent, Stan really only opens up when engaged in discussion of his life's passion – forest ecology and the enigma of hemlock in particular. Stan quickly acquired a reputation for eccentricity; frequently, for example, he could be seen sporting shorts in November. But for his work there has been nothing but respect, for Stan figured out why hemlock was failing in Algonquin Park.

Many "field" biologists these days actually spend little time in the "field," being seemingly cemented to their computer screens. Not Stan; he seemed a one-man renaissance of the days when wildlife science was less complicated and biologists were really professional field naturalists. Stan was rarely seen in the office, roaming the hills and vales of the Park for days on end, and becoming well-known to the many students and rangers who work in the interior of the Park. One wag coined the epithet "Sherlock Hemlock" and it stuck (the less flattering "dome gnome" being simply not fair, as Stan is not any shorter than most of us). Something of a bull in a china shop, he had a penchant for getting trucks stuck in the most remote of places and having to be rescued. In one such episode during a particularly wet spring, a Jeep was sent out only to get stuck itself (quite a feat, actually, for any 4WD vehicle) and then a full-blown wrecker which, incredibly, got stuck in turn. It took days, and the onset of a dry spell, to sort out the fiasco.

Stan, it is suspected, deduced the answer to the hemlock riddle before he started but, being a good scientist, approached the question with an open mind. He entered into the Park's forests and, with dogged determination, measured and cored thousands of hemlock trees (coring is the process of boring a section of wood to the tree's centre and extracting it to count growth rings). Before long, one thing had become abundantly clear: hemlock were reproducing, profusely in fact, but the tiny seedlings were not surviving. Looking at the numbers of hemlock trees in each age group, it was obvious that there were plenty of seedlings and no serious lack of the middle-aged trees and old giants but nothing in between.[36] Stan noted that the missing trees – the saplings and "polewood" – would all have started life during the period in history (around 1880 to the present) that deer (and subsequently moose) proliferated in the Park. The suspicion grew that deer and moose were the cause of the problem. Were deer and moose eating hemlock and thus destroying the very tree that they most rely on?

To test his theory, Stan marked hundreds of seedlings and transplanted others to areas frequented by moose (recall that deer are now scarce in the Park in winter). Moose and deer leave a telltale rough edge on a browsed twig because, unlike other browsers like hares, they lack upper front teeth and "tear" rather than bite. Stan could thus be sure of what he suspected was doing the damage. Upon returning the following spring, Stan had the answer he had long suspected; fully half of the transplanted, and a substantial portion of the marked seedlings, had been browsed by moose. Moose were (and still are) slowly but surely eliminating hemlock from the Park.

So, moose are removing hemlock, the essential cover for deer, which carry a parasite that is fatal to moose and it all depends on weather, logging, wolves and who knows what else in an elaborate ecological stew. Present-day moose may, unthinkingly, be doing their descendants a great favour by eliminating the favoured cover tree of their nemesis (deer).

A moment's thought, however, prompts an intriguing notion: if moose can so effortlessly eradicate young hemlock, and hemlock is a common tree in the forest, there must have been a time in the not-too-distant past when there were no, or very few, moose in the Park. Perhaps Cartier's "cerfz" and Champlain's "elk" really were elk or caribou and not moose. Or was there an incursion of deer into the Park about the time Columbus arrived (the oldest Hemlock in the Park being about 500 years)

*A winter's lunch.*

that brought brainworm with it that wiped out moose? Hemlock does fluctuate rather sharply in pollen cores and so must have gone through periods of boom and bust. Botanists who study the prehistory of trees, in fact, talk about a great die-off of hemlock throughout eastern North America about 5,000 years ago.[37] Did moose make a great thrust southwards? (To further muddy the waters, biologists in Pennsylvania determined long ago that deer also have a great fondness for hemlock as food, making one wonder how the tree endures at all.) There must have been periods when both deer and moose were relatively scarce. The whole subject is grist for some fascinating, but rather moot, speculation.

The story of moose and deer in Algonquin, having several episodes over the past 150-odd years, may be entering another chapter for it appears now that moose may be declining and deer recovering. Recent aerial surveys show that the moose population has declined rather sharply, while deer are being seen more and more frequently. There are even a few pockets of deer wintering in the Park now. Perhaps with the gradual onset of a warmer climate these trends will continue and deer will recover (although not likely to their former glory) as moose diminish. Most would lament this development since the moose, as misshapen and awkward as it may be, has become a much-loved symbol of the Park.

The history of moose, deer and Algonquin's forests is a fascinating one, but nothing has brought history to life in the Park as vividly as an astounding piece of work done in 1995 by two enterprising foresters, Bill Cole and Richard Guyette. The two, quite literally, have given new meaning to the term "Natural History."

The discovery occurred in the midst of an explosion of interest in the mid-1990s in learning how lakeshores work. A dedicated team of foresters and fisheries biologists focused intently for several years on discerning the fine details of how lakes and their surrounding forests "communicate" with each other (in an ecological sense). The group worked out of the Swan Lake Forest Reserve on the Park's west side, dissecting every detail of the reserve's forests and lakes for several years. Scott Lake, a small trout lake in the reserve, came to resemble one of those garish little vanilla doughnuts sprinkled with candy bits as its shoreline was festooned with multi-coloured ribbon and paint marking the location of hundreds of plots, probes and other peculiar devices of science.

Much of the outcome of this effort was rather humdrum: things like nutrient deposition rates, heavy metal loads, and the like – the fodder of scientific journals. Some stuff, however, was positively amazing. It was discovered, for example, that salamanders, by virtue of their reproductive cycle, essentially "feed" or fertilize our northern lakes.

Each spring, just after ice-out, thousands of yellow-spotted salamanders descend into lakes from the nearby uplands to lay eggs. Salamanders are something you rarely see because these amphibians spend almost their entire lives hidden under the dead leaves and rotting logs that litter the forest floor. It turns out, however, that they can be incredibly abundant; red-backed salamanders, for example, can number close to 30,000 per hectare (about 12,000 per acre) in deciduous forest![38] In fact, in terms of animal "biomass," or weight of living material, the forest salamanders far outweigh things like moose which, individually, of course, are much larger, but lack for numbers. Salamanders lay fist-sized gelatinous egg masses among the woody debris of the inshore shallows of lakes. Fish and practically everything else in the lakes love to eat these eggs and the larvae that hatch from them. The annual invasion of the amorous amphibians brings huge quantities of food, in the form of these eggs, into the lakes (2.4 million kilocalories per year into Scott Lake to be exact,[39] enough to grow 613 kg of brook trout). Salamander eggs are, in fact, one of the biggest sources of energy to the "food web," at the top of which sits its thriving trout fishery.

The same woody debris that is so loved by salamanders and fish but has filched the lures of generations of frustrated fisherman was the source of Richard Guyette and Bill Cole's great inspiration. Actually, it was an inquisitive fisheries biologist, Mark Ridgway, who first hit upon the idea of looking at submerged wood. Ridgway, director of the Park's Harkness Laboratory of Fisheries Research, became curious one day while wandering the shore of Swan Lake as to how old some of the larger logs really were. Ridgway was aware of the importance of submerged wood as a source of food and as cover to fish, and wondered about the "turn-around" time of the material – how long, if removed, it would take for the great mass of debris to rebuild naturally? This was no trivial question, for in the process of developing shorelines to build hotels, resorts and such things, the inconvenient debris is often pulled out to make way for beaches and boathouses – a good thing for swimmers, perhaps, but very hard on fish.

Ridgway approached Guyette and Cole with his question and the two, being foresters and used to thinking long-term, were immediately intrigued by the idea. Guyette and Cole had "cored" and aged thousands of trees on land but had never thought of looking at logs under water. They set about the task with a great fervor, adapting a 3,000-kg hand winch to haul the massive logs out of the lake and saw off sections for coring (focusing mainly on white pine, the most common species among the submerged and floating logs). They must, in doing so, have made for an odd spectacle: two ostensibly sane-looking men, struggling feverishly on shore with great masses of wood. Passing fishermen no doubt wondered if these were colleagues taking mad vengeance for the past thievery of lures. Over the space of a summer the two managed to remove, core (and replace) nearly all the old pine logs in Swan Lake.

It was in the lab that they discovered the true import of their labours. Cross sections of logs, it is well-known, reveal their age from counts of annual growth rings. Guyette and Cole, however, were not principally interested in how old the trees were but, rather, how long they had been dead (and fallen in the water). Foresters have devised an ingenious trick to tell how long logs have been down by making use of the imprint that time and weather leave in the wood. The technique is simple. A sound and very big live tree is selected and cut down and the pattern or "signature" of its growth rings is analyzed. The rings are, of course, laid down in summer when the tree grows, and represent the annual growth of wood at the periphery of the trunk. Every growing season is unique and the rings are never exactly the same width, varying with summer weather. The vagaries of summer weather thus form patterns in the growth rings that, from a newly felled tree, can be backdated in time from the most recent ring. These patterns of boom and bust are unique fingerprints and reflected identically in all nearby trees that grew during any part of the tenure of the "signature" tree. From "benchmarks" of these patterns, for example, an unusual prolonged growth spurt in the 1700s, a point in time can be established on any tree and then the date of birth of the tree established by counting back from that time.

By this means, starting with the youngest logs and building the archive sequentially back through time, Guyette and Cole were able to establish the germination date of every tree that had eventually became a log in Swan Lake. They were dumbfounded by their results, for the watery logs turned out to be centuries old. The oldest log had started as a seedling in about 900 AD, just as the Vikings started to raid England. It had died and fallen into the lake to become fish habitat 200 years later as the Crusaders entered Jerusalem. Most of the other logs had been in the lake for at least 400 years.[40] (The youngest log was dropped in 1890, possibly the result of the inexpert aim of an early logger.) Time, it seems, had stood still in Swan Lake. The cold, oxygen-poor water and resins in the pinewood had preserved the wood, creating a sort of tree mausoleum.

The import of this was staggering, because the entire shoreline ecosystem, the food factory of the lake, was built on material that had taken over a thousand years to accumulate. For nature to recreate this knotty, elaborate assembly, forged for a millennium, would, for all practical purposes, be impossible. Guyette himself has pointed out that if cottagers knew they had a thousand-year-old log in front of their summer home they might think twice about dragging it out and chopping it up for firewood!

There is another intriguing implication of Guyette and Cole's work. It is just possible that the water held in the core of these ancient logs is the same water that was in the trees when they fell hundreds of years ago. (The core of the logs is intact and removed from the periphery by 30 to 40 cm of resinous wood.) If so, the water is priceless, a relict piece of the biosphere dating from medieval time. The possibility opens up all sorts of avenues for scientific exploration, most obviously in the realm of acid rain.

I was at Swan Lake one day when Cole and Guyette brought up this possibility. We were in the station's makeshift lab and Rich Guyette was pressing a block of wood from the core of an old pine in a vice, squeezing out some of the precious water, all the while explaining the concept in an almost reverential tone. I was, of course, fascinated, but am ashamed to say that my entrepreneurial side was focused on the prospects for profit of making 1,000-year-old whiskey!

History was recreated in Algonquin Park in another ingenious way with a study of songbirds in the mid-'90s. One of the clarion calls of the conservation movement in the later part of the last century was the apparent widespread decline of migrating songbirds. Researchers first noted an apparent decline in "neotropical migrants" (songbirds that winter in the tropical parts of the Americas) in the late 1980s.[41] Suspicion fell immediately on the notorious deforestation of tropical rainforests as the cause. The problem with this theory, however, was that songbirds could be declining for a lot of other reasons; forests had also, of course, been "fragmented" on the breeding grounds in the U.S. and Canada, and even things like the profusion of domestic cats was taking a serious toll. What was needed was a means of isolating the various potential causes.

An opportunity to do so came about as a result of a long-forgotten piece of work that had been done with birds out of the Wildlife Station. In the mid-1950s, a researcher and naturalist with the rather imposing name of Norman Duncan Martin studied the forest habitat and songbirds in the general vicinity of the station.[42] Any good scientist takes pains to describe his methods accurately (recall the rigid structure of your own school lab reports – Purpose, Methods, Results, etc.) and Martin had done a superb job of recording his. The location of his study plots was particularly well-documented.

In 1994, Andrew Smith, a bright young graduate student from the University of Toronto, hit upon an ingenious idea, the substance of which was as follows. Songbirds, it seems, are declining, but we don't know if it is because of changes on the wintering grounds in South America or the breeding grounds in North America. Suppose there had been an accurate survey of songbirds somewhere in the U.S. or Canada *before* the declines started. If it could be shown that the habitat at this location had not changed appreciably yet the bird populations had declined, then the causal agent *must* be acting on the wintering grounds (conditions at the breeding site having remained static). The birds in question are essentially all forest birds and Smith needed a protected area, a Park, where the forests had been in or near a "climax," or stable state, for the last 40 or so years. He scoured the literature and came up with Martin's study which had been done in forest stands around the station that had been relatively undisturbed in the '50s and untouched since. Smith then searched out Norm Martin (who was retired but still active) and Martin, with the aid of his original papers and memory, was able to locate his study sites precisely.

Smith then spent the summers of 1995 and 1996 precisely repeating Martin's original surveys. The forest, of course, had changed but only in subtle ways that were not likely discernible to birds. To his surprise, neither had the birds. In fact, although there were slight changes in abundance of some species (ovenbirds, for example, had decreased) there were no significant changes in any of 41 species of birds in 43 years (including seven species that don't migrate).[43] Most of the migrating birds had actually increased.

Andrew Smith's work, in itself, certainly does not refute the entire rainforest/ declining bird scenario but does give one pause for thought. If songbirds are declining due to the changes on their wintering grounds, we should see it in places like Algonquin Park. The whole issue of declining songbirds is, in fact, far from (no pun intended) clear-cut. There is no doubt that some species, particularly those that migrate great distances, have declined, but some studies suggest that most songbird populations have not changed at all.[44] Smith's work demonstrates the immense value of having large protected area like Algonquin Park to do such research.

# 4 | Of Tooth and Claw

*The one common propensity of animals is to live if they can and die if they must.*[1]
— Paul Errington, 1974.

Paul Errington was something of a predator himself. Errington lived from 1902 to 1962 and fashioned his life's work with dogged resolve into perhaps the greatest single achievement in wildlife science. He started out, however, as a modest trapper, working the swales and potholes of South Dakota as a young man for thirteen years, in search, principally, of the equally modest animal that would bring him great fame – the muskrat.

Errington had developed an intense fascination with nature, and predators in particular, from childhood. He spent days as a young boy wandering the pastures of his family's Dakota farm and, near his death, wrote about an encounter he had during one such sojourn with a dying sheep. The sheep had had its side eaten off by a coyote and Erringtom noted dryly that afterwards, he; "did not feel that he would ever want to go out alone in the countryside."[2] The feeling was, fortunately for wildlife science, soon forgotten. Errington held on to his love of the outdoors, took up trapping and, fascinated by what he saw of violent death and its consequences in nature, decided to apply his meagre savings to an education in biology. Paul Errington went on to towering achievements in academia, developing the first comprehensive model to account for an enigma that is perhaps the overriding obsession of wildlife science – predation.

The very word predator conjures up feral images of jaws, teeth and bloody death. We have inherited a powerful, primal fear of predators from our Neolithic ancestors, an atavistic fear that is resurrected in the nightmares of childhood or found in unguarded moments alone in the dark. Man's foremost interest in wildlife seems to be centred on predators; witness the popularity of TV "science" programming which deals almost exclusively with stories of sharks, bears and lions or tigers. Predators and predation have an elementary appeal to biologists that was captured splendidly, by Errington himself:

> "To a biologist the concept of Death being a part of Life may seem logical enough. Predation, a way of Life resulting in the death of animals predated upon, is about as logical a consequence of Life's faculty for exploiting Life as anything that happens in the natural relationships of living things"[3]

This fascination with predators has accounted for a lopsided share of wildlife research, much of which has centred on the very basic question of whether predators limit or "control" prey populations. Algonquin Park, as we shall see, has carried its share of these studies but before looking at Algonquin, let's return to Errington himself, for the man, and his epic inquiry into the life cycle of the lowly muskrat, provided the first coherent view of the most thorny and controversial question in wildlife science: do predators control the numbers of their prey?

There is an intriguing photograph of Paul Errington on the jacket of his seminal book, *Muskrat Populations*, published in early '60s. The faded black and white photo shows a late middle-aged man deftly paddling a canoe through the thick morass of an Iowa cattail marsh. Errington is looking directly at the camera, squinting myopically in the sunshine through bottle-thick glasses. The eyes have a curiously oriental look and one can sense the intelligence in the commanding stare. More telling, though, is the manner in which he is handling the canoe. Errington's hands are positioned high and low and the paddle canted in execution of the classic "J" stroke. The image fairly bursts with energy and skill, and not surprisingly, for Errington, the quintessential "field man," spent a lifetime exploring the marshes, sloughs and open prairie of the Midwest.

Biologists naturally gravitate to the study of the "glamour" species – sharks, bears and the like, animals some of us refer to scornfully as the "charismatic megafauna." Therein lies the advantages of star appeal and often access to large research grants (and, let's face it, such things are just plain *exciting*). Errington, though, possibly with the Midwesterner's inclination to practicality, chose to study the lowly muskrat. He could not possibly have selected a more unpretentious subject.

Muskrats are nondescript brownish rodents about the size of your foot with a longish tail that they use to maneuver adroitly through cattail swamps, their preferred habitat. Exceedingly common, muskrats range from Alaska to Louisiana and Newfoundland to California but are more numerous in nutrient-rich swamplands than the crystal-clear but poor wetlands we have in the Park. The wretched rodents can't even claim the limited distinction of being rats, being in fact a type of (albeit very large) mouse, or more properly a vole.

Muskrats, like beaver, have long been a mainstay of the fur industry; many a young farm boy has made pocket change through school trapping and skinning "rats." The animals build and inhabit conical houses of vegetation that stick prominently two or three feet above the surface of swamps. This, and their foraging habits, leave ample evidence of their presence, something Errington took great advantage of. Muskrat populations are prone to wild fluctuations and are well-known to country folk for their great die-offs when they are commonly seen staggering across, and expiring on, lawns and fields. Errington set out to find out why.

Swamps are rich places and full of life but can be exceptionally tough to work in. Errington spent thousands of hours – a lifetime – exploring the swamps and sloughs around Ames, Iowa, making meticulous record of every fine detail of muskrat life. His journals make fascinating reading as one traces the man's developing expertise in the art of, in his own words, "the reading of sign." Errington did not often see his muskrats, except as corpses, but rather used the "sign" of their presence as his polling device. He also developed various ingenious and even comical ruses to make muskrats reveal their presence.

One such manoeuvre took advantage of the fact that in early winter of some years swamps may freeze before snowfall and one can walk the surface freely, observing every detail of animal life through the glassy ice of what is essentially a gigantic aquarium. Errington developed the habit in such conditions of stealthily approaching, then beating furiously on a muskrat lodge while singing aloud at the top of his lungs, forcing the astonished family of muskrats to flee, thereby facilitating an easy count through the ice. Errington "censused" entire marshes in this manner, developing means of exposing the more recalcitrant rats (who presumably had become habituated to his antics) and thus ensuring a total count.[4] The man, you understand, was a world-class scientist, and went on to great honours including a Fellowship in the American Association of the Advancement of Science. There is, however, something irregular, even disturbing, about a grown man prancing maniacally around a swamp in a the bitter cold of a northern prairie November. One wonders that he was not suspected as a lunatic.

Errington essentially saturated the swamps with effort and came to understand his subjects through sheer force of will. He came soon to realize that muskrats are subject to terrible periodic scourges of disease during which they leave their lodges in great distress, dying, or soon to die from the attentions of various predators in waiting, or other muskrats. In the "reading of sign," no news, a quiet marsh, was good news or, as Errington put it, "the scarcity of external tracks around an Iowa marsh where a substantial population of muskrats is known to be wintering may be indication that the animals are getting along well; and the converse almost inevitably is an indication of something being wrong."[5]

Muskrats, as goes the cliché, lead "short and brutish" lives and the violent outbreaks of disease and its consequences make it so. Errington soon discovered that the disease worked by causing massive internal bleeding and so called it the "hemorrhagic disease." He was constantly on the lookout for the first signs of an outbreak, dead muskrats or "floaters," the putrid condition of which he described dryly as often "being so rotten as to lose hair and appendages but their viscera may still be sufficiently intact to show distinguishable disease lesions in a fair proportion of cases."[6] Errington dissected and examined in great detail hundreds of these "floaters."

Errington, in fact, became a great chronicler of muskrat misery and it is worth drawing a long quote from *Muskrat Populations* because it captures brilliantly the brutish reality of muskrat life, a reality that, in variation, is typical of the experience of life of most wildlife:

> "Concerning the hundreds of muskrats examined as winter mortality victims, whether killed or fed upon by minks or not, the following may be said: They included trap cripples, with stumps from the wringing-off inflamed or cleanly healed. Victims' bodies were free of other wounds or with all parts bitten by other muskrats. Victims were found with fragmentary remains but with partly bare tail vertebrae from which the once-frozen flesh had been gnawed in life. They were individuals attaining sexual maturity and corresponding restlessness ahead of schedule. They were those breaking out of lodges or at the edge of the ice, to die outside of old or new injuries, hunger, exposure, disease, or by direct attack of mink, fox, dog, or other predator finding them at a disadvantage."[7]

Brutish indeed.

Errington never did figure out what actually caused the hemorrhagic disease, but it was almost certainly a virus. There is a large group of viruses that cause rapid, extensive and fatal hemorrhage in mammals, malicious agents that start the bleeding and then cruelly disable the chemicals that cause blood to clot and stop the bleeding. These include the dreaded and much-hyped Ebola of Africa that has been the inspiration for several bad movies. Errington did not know of Ebola but was, after all, a biologist and knew that by playing with the unknown he was courting danger. He let slip his feelings with a surreptitious passage buried amidst the exhaustive detail of the 700 pages of *Muskrat Populations,* stating poignantly that "I feel afraid of the disease."[8]

I got a sense of this sort of fear once while working on a study of lynx in Northern Ontario. Like Errington, we were endeavoring to understand the population biology of our subjects. Also like Errington, we were dissecting large numbers of carcasses, unearthing the secrets of the animal's lives from detailed examination of teeth, ovaries and stomach contents, etc. We dissected hundreds of lynxes donated by trappers and, with time, became quite efficient at it, working as a team in the lab in an assembly-line mode. I was normally, and unhappily, assigned the lynx's hind end and became quite adept at digging out the various parts found in those quarters (having done so to nearly 2,000 lynx I'm sure I must hold some sort of record). During one such session I got careless and took a nasty gash on my thumb from a sharp exposed claw. Ed Addison,[9] who is among many other things an expert on anatomy, was there that day to coach us through some techniques. Noticing my plight, Ed remarked gently that I had better hope not to get "Cat Scratch Fever." There was a good deal of blood and gore flying about, but I assumed he was joking and just chuckled and got back to work.

Some weeks later, while visiting our research labs north of Toronto, I looked Ed up for coffee. We were chatting idly until he suddenly became serious and asked if I had developed any signs of Cat Scratch Fever. I started to laugh until a glance at his eyes told me he was serious. Thoroughly alarmed, I listened to Ed (who is a wildlife disease specialist) describe a little-known but potentially serious disease that can be transmitted from cats (any cat) to man.

As is well known, cats, unlike dogs, retract their claws into sheathes in their paws when not in use. These sheaths collect blood and other offal and become perfect little reservoirs for a cocktail of germs. One of these, an unidentified bacterium, or possibly a virus of the type that causes herpes, can be transmitted to people through scratches that break the skin. The disease that develops is normally quite mild, with symptoms that doctors almost always mistake for the flu (Cat Scratch Fever may actually be much more common than generally accepted – 30 per cent of veterinarians, for example, show positive skin tests to the disease.) Although a rare occurrence, the illness can "progress" to involve the central nervous system with blindness, convulsions, coma and even death.[10]

I did not develop Cat Scratch Fever, but the whole episode gave me a new and guarded view of the animals. Henceforth, I have been exceedingly careful in handling cats of any sort, including the pompous little villain that we own.

With time, Errington came to realize that the hemorrhagic disease erupted when muskrat numbers reached a certain peak but lay dormant most of the time. Between outbreaks, the agent of the disease (again, presumably a virus) retreated to hide within the muskrat burrows. Muskrats were thus living in false security, surrounded by a lurking scourge. The disease seemed to emanate repeatedly from the very best muskrat territories and these choice spots were scattered about the marsh, forming a permanent gridwork of death. The "hot" zones could become so virulent as to be uninhabitable by muskrats for years – a spooky analogy to some of the toxic wastelands our nastier industries have created.

As long as the disease was dormant, muskrats were safe and secure in their burrows but, with an outbreak, the animals fled their erstwhile havens to certain doom. Errington knew that a lot of things killed muskrats: drought, floods, other muskrats and a host of predators including owls, snapping turtles and, above all, minks. He gradually, however, conceived of the momentous notion that all these things were irrelevant. The hemorrhagic disease so ruled the marsh that it, and it alone, "counted." Any muskrats surplus to a level set by the disease were condemned and it mattered not a whit what killed them.

Thus was born the concept of the "doomed surplus";[11] a large portion of the population was certain to die and the predation by turtles, minks and others was, in a sense, redundant. As Errington dryly put it "the accrued evidence indicates that much predation may operate in an incidental fashion rather than a true population depressant" and "in muskrats, the predation does not really count; it is centred on overproduced young muskrats and ailing and battered individuals."[12]

Not long after Errington died, another eminent researcher came forth with results of a major study that seemed to support the concept of the "doomed surplus."

Gordon Gullion spent a lifetime studying ruffed grouse, (our familiar "partridge") in the dense and tangled "bush" of northern Minnesota. Gullion, a professor of wildlife biology at the University of Minnesota in Cloquet, meticulously recorded the fate of grouse that he banded on mixed poplar-spruce woods for decades. He found that virtually all grouse are killed by predators and in particular by goshawks, a common hawk of the deep northern coniferous woods.

One might therefore be inclined to conclude that predation was important to the control of grouse numbers. Gullion, however, determined that the grouse killed by hawks were nearly all young birds forced out of prime habitats by aggressive territorial adults. Grouse that could find and hold good territories with lots of escape cover were fairly secure, those that could not were essentially condemned. Hawks weren't really killing the grouse; it was the grouse themselves. The parallel with Errington's muskrats was striking; here again was the "doomed surplus" although, in this case, the grouse set their own yardstick.[13] Predation, although present and very active, was, again, "irrelevant."

Errington and Gullion's work was so convincing and their reputations so formidable that they conditioned biologist's thinking about predation for decades. Predation just didn't matter.

The great beauty of science is its pure objectivity: the unsullied search for Truth. It must be said, however, that the teachings of Errington, Gullion and others were embraced perhaps a bit too eagerly by biologists of the time. The science of wildlife management was then in its infancy and the focus was on the very basics. There was a great deal of concern about the impact of hunting (which, with increased leisure time, was becoming hugely popular) on wildlife. By saying that predation didn't count, Errington seemed to, and in fact did, endorse hunting (Man the Predator).

Errington, it must be noted, never did say that predators were *always* irrelevant, pointing out himself that man the predator had entirely eradicated prey populations many times. His work, however, filled a sort of void and helped bring shape to the fundamental concept of "compensation." Compensation occurs when one type of death takes the place of another. In the case of hunting, if the hunter

*During the acute meat shortage of 1917, the Ontario government ordered that a number of deer be taken from Algonquin Park and shipped to Toronto to supply meat for the less fortunate of that city. Park Superintendent George Bartlett ordered the Rangers at Joe Lake to proceed with the chore. As the deer were brought from the bush to the railway tracks, Edwin Thomas picked them up and transported them to Joe Lake Station with the railway motor car. Only the largest deer were taken. Mr. and Mrs. Edwin Thomas pose beside two large bucks and one doe on the Joe Lake station platform, ready for shipment to Toronto.* Algonquin Park Museum Archives #31 – Jack Wilkinson.

*Deer hunters at Rock Lake.* Algonquin Park Museum Archives #6335 – Robert Taylor.

removes animals that would have died anyway of, say, disease, "compensation" has occurred and the hunting doesn't matter (to the population at least; it obviously matters to the animal removed). This simple concept formed the early framework of much of the science of wildlife management.

I do not mean to sound entirely negative about this, for hunting in fact does not normally limit wildlife populations. For proof, consider the fact that over three million white-tailed deer are taken by hunters every year in eastern North America and have been for decades,[14] yet deer populations are flourishing everywhere. Errington's work was nevertheless embraced perhaps a little too happily.

But few things in nature are simple, and all of this talk of "doomed surpluses" and "compensation" began to get a second look when biologists turned to the study of things bigger than the diminutive mink and muskrat. I'm talking about wolves.

Biologists began to look seriously at wolves in the late '50s and, this being a book about Algonquin Park, it's time we turned to the classic early study of the wolf – an extraordinary and historic piece of work that gave momentum to the early foundation of the Wildlife Station, and all the great things that followed.

Wolves have captured the imagination of man from primordial time down through the eons to the present day. We have inherited a deep-seated fear and prejudice of wolves from our prehistoric ancestors. These attitudes die very hard and, in spite of the growing ecological awareness of our times, many people still view wolves with a great and morbid dread. This is terribly unfortunate because wolves, with their rich and complex family life and marvellous pack-hunting skills, are among the most fascinating of wild animals. Wolves are also, incidentally, not particularly dangerous (at least to people). There has *never*, at least not in modern history, been a confirmed case of a fatal attack by wild wolves on a person in North America. Wolves and people, nevertheless, don't get along and wolves are for the most part consigned to exile in true wilderness – places like Algonquin Park, where they are, at least most of the time, at arms length from people.[15]

Algonquin wolves are an odd lot. Small (at most 90 pounds), slight of build, and with a tawny brownish colour indicative of coyotes, they have been the subject of much speculation by biologists, some suggesting they are not wolves at all but some sort of wolf-coyote hybrid. Very recent and fascinating research, however, suggests something very different. It seems that Algonquin's wolves are a relict population of the endangered southern Red Wolf, which originally ranged well south of the Park but moved north with deer following the clearing of the forests.[16] But if today's wolves are new arrivals, what sort of wolf existed in the Park in the first place? We know that today's Algonquin wolves, although they will kill moose, are not particularly effective moose predators, being much more effective at killing deer. This in itself suggests some sort of substitution has occurred because, if it is true that deer were not here prior to settlement and moose were,[17] the original Algonquin wolves were probably a larger moose-killing breed, more like the northern "Grey" Wolf.

Man, at any rate, has had a long-standing interest in wolves, and it is thus perhaps surprising that serious research on wolves did not get underway until much work on muskrats, mice and many other "lesser" things was done. In fact, no serious work was done on wolves in Canada (where almost all wolves in North America are found) until the '60s.

Then, along came Douglas Pimlott.

Doug Pimlott was a solidly built six-footer with a passion for wolves and the workaholic's drive to succeed. Like a lot of leading wildlifers of the time, he was a forester by training. An American by birth, Pimlott came to Canada after having served in the Navy in World War II and worked in Newfoundland before arriving in Algonquin in Park in 1958. The work he did with wolves in Algonquin from 1958 to 1963 made him one of the best-known wildlife researchers of his time. Pimlott was, like a lot of scientists, a quiet, reserved man eschewing, for example, the nightly poker games at the station for the tranquility of his cabin where he would read and write in peace into the wee hours. He was also a religious man who neither smoked nor drank nor swore (something that, as we shall see, he had ample cause to do during the course of his research). Apart from leaving the *Christian Science Monitor* in view on the coffee table in the station's cookhouse, he nevertheless kept his religious beliefs to himself. In spite of his natural reticence, Pimlott was a leader, his congenial, controlled nature and the example of his prodigious work habits bringing and keeping the wolf research team together.

He certainly had his family behind him. Bruce Stephenson, who was Director of the Wildlife Station through much of the '60s, recalls visiting Pimlott at his home north of Toronto after the wolf study was over. He found Pimlott upstairs in his study writing up the final report of the project. Pimlott was scribbling furiously on a writing pad with an awkward and very large script of three words to the line. His wife would periodically run up to retrieve the pages scattered on the floor and type while his daughter did the graphs and his son the tables.[18]

There being, at any rate, almost nothing known about eastern forest-dwelling wolves, Pimlott and his co-workers set about their work with a clean slate. In general, there was very little of substance known about wolves, and many doubts and questions about their effects as predators on prey.

This is not to say that prior to Pimlott's arrival the wolves of Algonquin Park were ignored. On the contrary, Algonquin's wolves had been subjected for decades to an earnest and concerted effort to "control," if not exterminate, them. From 1909 to 1957, Park Wardens

"The Wolf Hunters," a C.J. Bond photograph dated March 3, 1930. Shown left to right are Jack Millar and Jack Gervais. Courtesy of Algonquin Park Museum Archives #6793 – Don Beauprie.

poisoned, shot, snared and trapped wolves in a determined, if misguided, effort to protect the Park's deer (which were, unlike the wolves, viewed as having a certain intrinsic value). From 1955 to '57, for example, 220 wolves were poisoned or snared in the Park.[19]

Faded photographs from the '30s hang on various office walls in the Park displaying the men who did this, all posed theatrically in bush camps amidst the results of their work – piles of carcasses and pelts of wolves, beaver and, oddly, loons (the loonskins were used as winter moccasin liners). The images have the dirty brownish wash of photos of the time and show rugged, unshaven men dressed in thick flannel, sporting oversized badges on their puffed chests and struck in dramatic attitudes like great Bwanas of Africa, staring haughtily at the camera lens. The Wardens were proud, tough and independent fellows and yet they appear quaint, even comical, in the obviously contrived scenes. One wonders if they ever had second thoughts, if they ever questioned the wisdom of their butchery. From the look in their eyes I rather doubt it.

The Warden's efforts were, at any rate, in vain, for the wolves weathered this storm nicely, there being no clear evidence that the killing ever significantly reduced their numbers.[20] In all likelihood, the wolf population simply reacted to the killing by producing more pups – a classic "compensation" response.

Pimlott's young research team chose to study wolves in a roughly 3,000-square-kilometre area of rugged rocky hills, dense forest and clear headwater lakes centred around the newly established wildlife station. The area is a magnificent wilderness, in its own way without comparison, but was (and still is) almost totally inaccessible, and this presented formidable challenges.

Pimlott lacked a considerable advantage enjoyed by biologists today: the use of radio telemetry. Biologists nowadays employ the use of remote tracking devices (telemetry) almost as a matter of course. Some of the technology has become incredibly sophisticated, allowing researchers to study animals literally from their living rooms. There are, for example, systems that track and map animal movements automatically from satellites, and radio collars used on bears that can re-inject the animal with built-in tranquilizing darts and detach

*A rare photograph of Doug Pimlott, shown here with a wolf cub.* Algonquin Park Museum Archives – Mark Pimlott.

themselves by remote command. Pimlott and his co-workers had to study wolves basically with their feet.

Pimlott did have occasional use of a helicopter and was able to do a crude census of wolf packs in the study area by intensive low-level flights in winter. However, ground searches showed that the helicopter missed animals, particularly lone wolves (and was, anyway, useless in summer) so the team experimented with other techniques. Howling, which will often evoke a response from wolves in late summer, was tried but deemed unreliable (these early experiments, though, were the inspiration for Algonquin's famous Public Wolf Howls). One enterprising fellow hit upon the idea of enticing wolves to visit and pee on wooden posts scented with wolf urine (as dogs do on lampposts). The idea was to place the posts strategically about the study area and to count tracks of wolves visiting the posts. The wolves, however, scornfully ignored the posts, showing more interest in coloured flags placed nearby to help the researchers find them. Then moose began eating the flags and the project was promptly abandoned.

In the end there was nothing for it but to press on, and Pimlott's crew spent thousands of hours plowing through deep snow and slogging through swamps and bush in (no pun intended) "dogged" pursuit of the wolves. Much of this work was done at night, a hazardous venture in the Algonquin wilds, not because of the risk of attack by bear (which is slim) or wolves (which is even less likely) but rather the stony ground that can easily trap a boot and snap a leg in the dark.

All, however, was not work at the station, for even dedicated scientists have to have a bit of fun from time to time. A friendly baseball rivalry had developed over the years between the wildlifers at the station and the fisheries researchers at the Harkness Laboratory on Lake Opeongo. Results of some two decades of this sport are etched on a rough pine plaque that still hangs in a dusty corner of the cookhouse (the wildlifers, having honed their footwork chasing wolves and bears, generally prevailed). Baseball in Algonquin was simply a diversion for overtaxed minds and unremarkable but for a bizarre event that occurred one summer in the early '60s. The event involved Pimlott's wolves.

Games were played on an abandoned airfield behind the Lake of Two Rivers campground. Halfway through one summer's season, it was noticed that the bases, crude squares of plywood, were disappearing mysteriously from game to game. It was assumed that some knave from the nearby campgrounds was (again, no pun intended) "stealing" them, presumably for firewood. This became the widely (if skeptically) held explanation for the mystery and the bases were simply renewed every game, until a player arriving early one evening caught a glimpse of a wolf bounding into the woods bordering the field with a base in its mouth! Incredulity reigned, but Pimlott was called in and an inspection of sign and tracks revealed, sure enough, that a wolf was making off with the bases.

The story would have remained a quaint anecdote but for the unfortunate coincidence that rabies had just been confirmed in a wolf for the first time in North America. This prompted a conference with Park authorities and it was decided that the behaviour of the wolf was so aberrant, and the risk to nearby campers real enough, that the animal had to be shot. So it was, and, not surprisingly (the disease being extremely rare outside of foxes, skunks and raccoons), the wolf tested negative for rabies.

Baseball was not the only way that the rivalry between the Station and Harkness was played out. There was the annual Frog Jumping Contest, an event that entailed a great deal of wagering and drinking of beer. In mid-summer every year, each side would choose a frog and meet, by custom, at an agreed upon neutral site. The frogs would be marked and placed under a basket around which concentric circles had been drawn, the winner being, of course, the frog to first cross the outermost circle. One year, the Wildlife Station happened to have a frog expert on hand and he was, naturally, given the task of choosing the frog. The fisheries team, on the other hand, had completely forgotten the event and (the truth was never made clear) either caught their representative in the rush over or accepted one from the station staff. Both frogs were placed under the basket which, after the appropriate delay, was lifted – to reveal only one frog. The Wildlife Station's entry was declared the unanimous winner when the smaller frog's hind feet were seen dangling from its mouth. The herpetologist, it was later admitted, had chosen a predacious frog of a variety unknown to the fisheries people.[21]

With such episodes providing for much needed comic relief, Pimlott and his team soldiered on. In time, a basic understanding of Algonquin's wolves began to emerge. Pimlott came to estimate that there were about 300 wolves in the Park found in 40 to 50 packs. This was the first credible estimate of a wolf population east of Minnesota and is considered today to have been remarkably accurate, given the technology of the era.

Pimlott also took a shot at estimating the number of moose in the Park. To do so, he did counts from low-flying aircraft in winter, a technique that, with only fine differences in design, is widely employed today. Moose, being big and black, are easily visible against the snow in winter and, under the right conditions, can be censused quite reliably. Biologists across North America now routinely survey moose on standard 25-square-kilometre plots and apply the average number of moose per plot to estimate entire populations. Pimlott's estimate of

*George Phillips, a member of the Canadian Aviation Hall of Fame, was one of Algonquin Park's flying superintendents from 1944 to 1962. Others included the colourful Yorkie Fisker and the very popular Frank MacDougall. Aircraft were used in the earliest wildlife surveys, including Pimlott's work.* Ontario Archives, Publisher's Collection.

*An aerial view of Lake Opeongo, Ted Jenkins, photographer.* Ministry of Natural Resources, Publisher's Collection.

about 1,500 moose for the Park in the late '50s[22] would probably have been duplicated by today's technique.

The work, flying at near treetop level on the bright (but bitter cold) days of mid-winter when moose are out "sunning," is inherently dangerous. Aircraft go down on wildlife surveys all too frequently. But no incident I know of matches, for sheer weirdness, a crash that occurred in the Chapleau district of Ontario one winter in the early '90s.

The event occurred on a routine survey involving a specialized short-takeoff aircraft known as a Maule M7 and a survey crew of three (pilot, navigator and one observer). When moose are spotted it is customary to "buzz" them – that is, swoop down for a close look to determine their sex and a rough estimate of age (and also, truth be known, for the fun of it). This crew on this day had spotted a thick concentration of moose tracks near the start of their last flight line and swooped low to investigate. The observer spotted a bull moose out of the corner of his eye and yelled for the pilot to turn. This he did, making a sharp bank to the left that was perhaps a bit too steep, for they shortly found themselves flying below the treetops. The pilot powered up but not soon enough and the plane hit the trees hard, tearing off a wing. The result of this was that the aircraft turned turtle into the spruce trees and deep snow. One advantage of flying in light-bodied aircraft is that the impact of a crash is often relatively slight and, cushioned by the trees and snow, the plane came down gently. The crew emerged shaken, cut and bruised but, apart from a suspected broken arm, not seriously hurt. Gathering about to discuss their next move, they were shocked to see the plane rocking gently from side to side. Then, incredibly, a bull moose surfaced like a breaching whale from beneath the twisted wreckage. The plane had landed right on the moose! The moose gave (to their brief consternation) a long hard look their way and then snorted, and trotted away disdainfully, to all appearances unscathed.[23]

Aerial surveys are dangerous and a great many biologists have been killed in plane crashes. Bush pilots are notoriously cavalier. A colleague of mine went down in a helicopter over the James Bay Lowlands while doing a beaver survey because the machine *ran out of gas*. The fuel gauge, it seems, wasn't working and the pilot had gotten into the habit of checking the gas with a wooden stick every morning.

On this particular day he had simply forgotten. Here again, no one was hurt in part because helicopters, if the engine gears can be disengaged, will "autorotate," the blades turning from the draft of descent and cushioning the impact. Another colleague who survived a similar event recalls, vividly, the insides of the chopper afterwards having a glaze of vomit, the product of the violent spinning and sheer terror.

All this effort is about *numbers*, knowing the size of populations being, of course, pivotal to understanding and "managing" them. Biologists are obsessed with numbers and have devised a myriad of clever ways to survey populations, some being very high tech. Often, though, it comes down to just getting out in the bush and counting. Paul Errington, being a down-to-earth fellow, preferred it that way and left us with this beautiful account of his unadorned approach to the problem:

> "My own procedure in estimating numbers of muskrats from signs has been to consider the signs about each lodge or burrow system, then think back to my trapping experience and estimate that I could have expected to catch at least so many muskrats and no more than so many at a particular place. The two figures would be put down in my field notes. Then, I would put a check mark beside the figure that I considered nearest the truth. In arriving at an estimate for the whole marsh or stretch of stream, I would add up the minima and maxima to get the range of estimates and, finally, in a separate column, add up the checked figures. The latter would be as close as I felt able to come to the true population through estimates."[24]

This is science at its simplest and yet most elegant.

Deer, unfortunately, cannot be counted from the air in winter (they hide under the evergreens) but, not to be deterred, biologists have figured out ways. The principal technique involves, I'm embarrassed to say, counting deer s__ – or "pellet groups." (Biologists, as we shall see, seem to have an unhealthy obsession with such stuff.) Someone figured out, presumably by watching, that deer defecate *exactly* 14.8 times per day.[25] That someone, I suggest, needs to get a life, but the observation provides the means to do a reasonably accurate count of deer. Quite simply, if you know the number of deer droppings on an area of land and the rate at which the deer defecate and the time they have been there you can, through simple math, track back to the number of deer. People have thus spent entire careers out in the bush counting deer crap.

I was involved in doing just that in the vicinity of Algonquin Park in the spring of 1975. The technique required laying out elongated plots and then carefully walking, or better yet, crawling up one side of the plot counting droppings within a one-metre zone, and then reversing direction and doing the same back to the starting point. We were doing so for a while near the Parry Sound District, a portion of which contains most of Ontario's limited population of rattlesnakes. My partner, who was a dedicated fellow, insisted upon squatting and awkwardly duck-walking his way up and down the plot, the better, he insisted, to get an accurate count. One time, returning down the second leg of a plot, he happened to glance at the area he had just covered and was shocked to see a fat rattlesnake coiled up at a point that had, moments before, placed it within inches of his butt!

But, to return once again to Algonquin Park and Doug Pimlott.

The Park was a lot different in Pimlott's time. The forest, as we have heard, was younger and more favourable for deer, and thus for wolves. Pimlott found that Algonquin's wolves were devoted deer killers; deer then comprised about 80 per cent of the diet, moose 9 per cent and beaver 7 per cent (the remainder being the odd rabbit or squirrel).[26] The wolves enjoyed the only real variety in this diet in summer, because that is when they could get at the beaver and also prey on highly vulnerable, and presumably very tasty, deer fawns. In winter, wolves had little else to eat but deer and preyed heavily on the snow-bound herds. (Moose were relatively scarce in the late '50s and most of the moose consumed was probably scavenged.)[27]

The pack hunting skills of Algonquin's wolves were most apparent in winter, when track patterns could be followed on snow. Pimlott's crew were fascinated with the hunting skills of wolves and Pimlott himself took an exceptional interest in these tactics, devoting much of his journals to accounts of wolf attacks on deer.

Wolf-killed deer are often seen out on frozen lakes in winter. Carcasses are, of course, more conspicuous out in the open, but this is not just a matter of visibility because wolves seem to kill a disproportionate number of deer on the ice. Biologists are not sure why. The hooves of deer are hard and blunt and useless on any slippery surface (try walking on ice with stilts). For at least part of the winter (before snow and slush get deep), deer are helpless on ice; their legs splay pathetically and can even break with the awkward torque, and they are vulnerable to wolves. The conditions that make for these easy pickings, though, are rare. We know that deer are often forced by deep snow into the pockets of conifer that fringe Algonquin's lakes, and it may simply be that wolves are killing deer near where they find them (as opposed to "driving" them onto the lakes).

Pimlott gives a rather clinical account of a what was typical of such encounters with the following:

> "On January 31 [1962] two wolves killed a large 10-year-old male deer on Smoke Lake .... The two wolves began to chase two deer just east of the Smoke Lake parking lot. The tracks of the deer and the wolves were parallel for about 200 yards; at this point the buck made a sharp turn toward the lake, then the smaller deer swung in the opposite direction and remained in the trees. The wolves caught the buck soon after he reached the lake and brought him down within 25 or 30 ft of their contact. They ate only a small amount before leaving the area. Passing cars may have disturbed the wolves as the kill was only a short distance off the highway."[28]

It is interesting that it was the old buck that was killed. Old bucks don't normally get that way by making mistakes.

It is also interesting that Pimlott took pains to note that the wolves ate only part of the carcass. Folklore around the Park, particularly

*Wolf and deer carcass.*

among veteran trappers, has it that wolves are wanton killers, frequently killing deer by scores and leaving the carcasses to waste. The wisdom of old trappers is something to be taken seriously, but Pimlott found very little evidence of such wastage. The research team found an incredible 676 deer carcasses that were believed to have been killed by wolves and, of these, two-thirds were entirely consumed when first found. Only about 10 per cent of carcasses were abandoned without being substantially eaten. (The very fact that Pimlott found 676 dead deer is a testament to the prodigious amount of footwork that went into the project.)

Such, then, was generally the pattern, wolves rarely wasted the deer they killed. But the winters 1959 and '60 were different. Those were particularly cruel winters for deer; snow conditions were severe and wolves were able to kill deer almost at will. Pimlott noticed that many more carcasses were being abandoned, presumably because the wolves were killing deer more quickly than they could eat them. Such

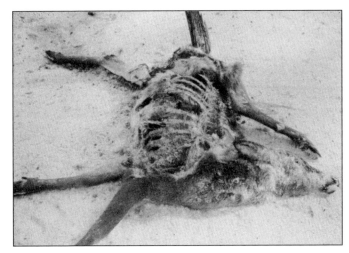

*A deer killed by wolves. Pimlott's team found hundreds of these.* Algonquin Park Museum Archives, MNR staff

conditions, though, are unusual, and Pimlott, clearly concerned about the issue, took great pains to point out that wolves were not wanton killers. (Subsequently, biologists have determined that when prey are really abundant, wolves may only consume 1/4 to 1/3 of their kill and then move on. Scavengers then move in.)

The persistent disagreement with trappers was nevertheless troubling to Pimlott, for he knew well that old trappers don't lie (although they may, in my experience, stretch the truth a bit). The confusion may have come from trappers forcing wolves off carcasses before the wolves finished eating. They may also have had stronger memories of the really tough winters.

Pimlott remained frustrated by the issue and, being an assiduous man, came up (perhaps a bit hastily) with a novel solution – some deer just don't taste good to wolves. This idea arose from some observations Pimlott and a colleague, Jack Shannon, made while following wolves to deer kills in the winter of '59. Shannon noticed that some of the deer carcasses abandoned by wolves had a "strong odour of the flesh" (one would expect they all did but everything, as they say, is relative).

Pimlott illustrated this with the account of an incident from March of '59:

> "Shannon located two deer that had been killed within a short period by the same pack. A 4 1/2-year-old adult male had been killed and then apparently had

been rejected, as it remained virtually untouched. The wolves left the site and killed a fawn only a short distance away. Approximately 66% of the latter had been consumed when Shannon arrived at the scene. While removing the femurs[29] of the two animals he noted that the flesh of the adult had a very strong odour, somewhat reminiscent of conifer oil in contrast to that of the fawn which presented the usual "sweet smelling" odour of fresh meat."[30]

Thus was born the "differential palatability hypothesis" (some deer just taste better than others). I suspect this, however, to be a rare departure from objectivity. Anyone who has seen a pack of wolves tuck into a deer will have trouble considering them fastidious. (Wolves can certainly pack it away. A single wolf was known to consume half of a small caribou carcass – 33 kg of meat – in less than 42 hours on Baffin Island!)[31]

Trappers are a contrary, headstrong lot and given to eccentricities. The truth of this was never better bespoken than by a wizened old veteran that used to appear frequently in our field office to have his fur (caught outside the Park) stamped for sale – more frequently, I recall, than necessary, because the poor chap seemed often to visit for the sole purpose of a good chat (it being lonely on the trapline). His favourite story was that of his lost thumb.

The fellow, it seems, was out ice fishing (not in the Park, of course, that is illegal) with one of the new power augers. A power auger is an ingenious device that uses a 2-cycle motor to bore holes through the ice with great speed. He had got his gloves wet and, it being bitter cold, they had frozen solidly to the handles of the auger. While attempting to cut a hole, the auger jammed and spun violently, pulling one glove away with such force as to tear the man's thumb off. All that remained was a hideous bony claw. Being a trapper, and tough as nails, the fellow calmly packed up (not forgetting his fish!) and dutifully delivered himself and thumb to the local hospital. The thumb, unfortunately, was beyond salvage and the man was packed off with

*The late legendary Ralph Bice, an outdoorsman, guide and trapper, shown out on the trapline with his sleigh dog, Max.*[32] Ontario Archives, Publisher's Collection.

great haste to Toronto for emergency reconstructive surgery. The surgeons chose, perversely, to take skin from the man's buttocks to graft onto the remnant of bone. I say perversely, because the fellow had an exceptionally hairy a – and the end product was a grotesque bright red, hairy knob that the gentleman never failed to display while relating the story with great glee. He always ended by announcing loudly that his "arse was itchy" and, giggling wickedly, scratching his thumb.

Wolves can exact a terrible toll on their victims but don't always have things their way, occasionally being the victims of predation themselves. There seems to be no love lost between wolves and bears, and clashes, although rare, are not uncommon. A student of Pimlott's, for example found evidence that a black bear (Algonquin's only variety) killed a female wolf right beside her den of pups in the summer of 1961.[33]

Wolves, though, generally get the better of these encounters.

In February of 1996, we were flying a moose survey in the vicinity of Turquoise Lake on the Park's east side. Moose surveys are always flown on sunny days to allow for the sharp contrast of shadows against the snow to help locate the animals. Around noon, a colleague caught a glimpse of a dark smudge against the sparkling bright snowy surface of the lake. Thinking it an otter, which love to skate across frozen lakes, we turned to have a look and saw three wolves emerge from the shoreline and circle the object (which we had by now surmised to be something dead). We landed to investigate, still expecting to find an otter, which occasionally fall prey to wolves on the bare ice. The wolves dispersed and upon reaching the mysterious black spot we were astonished to see that it was a bear (bears are supposed to be snug in their dens in mid-February).

It had snowed the day before, covering any tracks that might have revealed events, but there was an enticing clue to the mystery. The bear, last year's cub, had no pads on its feet. Bears, for reasons known only to bears, shed the pads of their feet while hibernating leaving soft, raw soles. This bear then, had come recently from a den (ruling out a host of possibilities such as having been shot by a poacher late in the previous fall). Wolves had nefariously entered the bear's den (which was presumably nearby), pulled the sleeping bear out, killed it and dragged it onto the lake. We looked around for the den or sign of mother bear or siblings, but had no luck and left the scene pondering the savage audacity of wolves.

Actually, it was not impossible that the bear had left the den and been caught out in the open. Bears will occasionally abandon their dens in mid-winter when flooded out by an early thaw. There had not, however, been a thaw for weeks.

The whole thing seemed terribly unfair; killing a sleeping baby bear in the den seemed beyond the pale even for natural brigands like wolves (honour among thieves and all). One must always be careful, however, not to judge the natural world and, fair or not, such attacks may not be that uncommon. David Mech, the renowned wolf expert, recorded a similar event in Minnesota in the winter of 1977. In this case, a pack of six wolves located and attacked a den containing a mother and two newborn cubs in mid-February (bear dens are often given away by air escaping from breathing holes and, in any case, must smell very plainly to wolves). The wolves were radio collared and Mech was on the scene right after the attack and able to record exactly what happened: "The wolves apparently attacked from both sides and drove the bear from the den. The bear fought her way 22 metres to the nearest big tree, a mature aspen, leaving a path of broken brush and bear fur. At the tree, the fight continued; trampled brush, part of a wolf canine tooth, tufts of wolf fur and much bear fur were concentrated in a 3 metre radius around the tree, but claw marks indicated that she climbed to the safety of the crown. She eventually came down and returned to the safety of the den where she died or was killed. Bear fur covered the snow within 2 or 3 metres of the den. Tracks visible in the photographs showed that the wolves dragged the carcass beyond the fur-covered area to consume it. By March 21, all that remained of the carcass were fur, fragments of bone, and the nearly intact skull. Wolf droppings in the vicinity contained claws of the newborn cubs."[34]

Honour indeed.

Very recently, biologists deciphered an equally gruesome event at a black bear den in Montana. In this case, the aggressor was a Grizzly Bear. Based on tracks, blood and a few pitiful remnants of bone and hair at the scene, the researchers determined that an adult and presumably very hungry Grizzly had entered the den of a young black bear in early winter, pulled the sleeping animal out, killed it and completely consumed it! The black bear had denned in a 12 metre diametre pile of logging debris the "roof" of which the Grizzly apparently ripped off to gain access. In investigating the scene, the biologists found the Grizzly dozing in a clump of spruce trees some six metres away. The enraged animal charged but was turned away by pepper spray at a distance of seven metres.[35]

Bears, at any rate, are marvellous animals but were only of marginal interest to Pimlott. He was interested in wolves and wolf populations. Pimlott was less concerned with deer themselves but greatly intrigued by the effects of wolves on deer. Deer were (despite the occasional bad winter) healthy and abundant through Pimlott's tenure in the Park. This was in spite of the relentless exertions of wolves which, Pimlott knew well, were killing hundreds of deer each year. Pimlott came to the fairly obvious conclusion that wolves were not limiting the deer population. He must have been influenced in this by Errington's thinking – the prevailing dogma of the time that predation didn't matter – but could not legitimately have concluded otherwise.

Yet Pimlott was troubled once again. He had observed first hand the rapacious efficiency of hunting wolves and had difficulty accepting that this fearful instrument was of no consequence. Searching for a solution, Pimlott conceived of the idea that perhaps the setting, the entire framework, of his experiment was flawed. Algonquin's forests, and indeed forests throughout eastern North America, were recovering from a period of intensive and uncontrolled commercial logging. The landscape, being mostly in "second growth" was excellent deer habitat, favouring the explosive growth of deer populations as long as winters weren't too severe. Pimlott knew that this situation was unnatural, that the native condition of our forests was not nearly so deer-friendly. Perhaps deer had an unfair advantage in the Algonquin of the 1950s. Perhaps their apparent success was an aberration, a sort of glitch resulting from what loggers had done to the landscape. Perhaps wolves really could suppress deer numbers in normal circumstances. Pimlott made the point in a brilliant paper written in 1967 in the *American Zoologist*.[36]

Pimlott had crossed the divide. Maybe predation did matter.

Doug Pimlott left Algonquin Park in the mid-1960s, just before the deer population crashed. The winters of 1969 and 1970 were, as we have seen earlier, brutal: deer died by thousands and the population has never really recovered. Pimlott was not around to witness the disaster but one of his students, Dennis Voigt, was.

Voigt, who, even today, retains boyish looks and a shock of bright blond hair had joined the Pimlott team as a young graduate student in the early '70s. He went on to a stellar career in field research, establishing a reputation as Ontario's premier expert on deer, but in the early '70s was focused on the events unfolding in Algonquin.

The deer had not just declined, they had all but disappeared. Yet wolves were still around the Park, scarcer perhaps, but still seen on occasion and, as always, heard howling in late summer. What, Voigt wondered, were they eating? Pimlott's work had shown Algonquin's wolves to be eaters of deer and there did not seem to be much out there in the way of alternatives.

Voigt fell back on the standard remedy of frustrated biologists: he collected scats (animal turds) – hundreds of them. The Pimlott team had actually been collecting wolf scats all along and Voigt extended the collection into the early '70s. Animal hairs have fine patterns of scaling on the surface, something like fingerprints, that are unique for each species (these patterns had been revealed by another of Pimlott's co-workers, George Kolenosky, in 1969). The scales, which form exquisite lattices visible under the microscope, remain intact even after passing through the digestive tract of a predator. Voigt (heedful, no doubt, of the dangers of hydatid disease) teased thousands of hairs out of hundreds of wolf scats and painstakingly reconstructed the eating habits of wolves in Algonquin for an entire decade (1963-72). Voigt found, essentially, that wolves had turned to beaver after the deer die-off; beaver went from being a minor component (7 per cent, see above) in the early 1960s to over half of the diet by 1972.[37] The wolves had simply "switched."

This, of course, was hard luck for the beaver, but the new-found attentions of wolves had not just diminished the beaver, it had changed their way of life. Alex Hall, another student inspired by Pimlott, studied beaver in and immediately outside of the Park in 1969-70. Hall found that beaver outside the Park (where wolves were presumably scarcer) fed heavily, as is their normal custom, on upland aspens and birch, well away from the security of their ponds. In the Park, where wolves were still common, beaver rarely strayed from their ponds, feeding almost exclusively on waterlilies. It had apparently become too dangerous for beaver in the Park to leave the water![38] (Pimlott himself had noticed that marauding wolves rarely passed a beaver lodge without checking it out). Beaver too had "switched."

Beaver had become the surrogate for deer but were of no use to wolves in winter while safe and snug in their lodges. Wolves had no choice in winter but to find the very few deer that remained. This they did with great skill for, although deer were, for all practical purposes, absent from the Park in the '70s, deer still formed a substantial part of the diet (up to 33 per cent in 1972, the last year of study). Wolves were able to root out the very few deer that remained. (Note that moose were still relatively scarce in the Park and formed only a minor item of wolf food – most of it probably being scavenged).

We see this amazing ability even today. Deer are still scarce in Algonquin in winter and we routinely pass by areas that show no sign of deer at all. Yet it is not uncommon for deer carcasses to mysteriously show up in these supposedly "deerless" spots – proof that what we cannot find, the wolves, in desperation, do.

The "prey switching" that Voigt and others saw in the Park was reflected some years later in a remarkable story that remains to this day one of the most extraordinary pieces of detective work in wildlife science. The events occurred on, of all places, the island of Newfoundland, but have striking parallels with Algonquin and thus a place in our narrative. The story, moreover, is simply amazing.

The affair centred on caribou. Caribou (otherwise known as reindeer) are a medium-sized member of the deer family, smaller than a moose but bigger than a deer. Caribou live almost exclusively on lichen, the yellowish-green stuff you see growing on rocks and, having this unpromising advantage, are superbly adapted to the windswept tundra of our far north. Newfoundland is nothing if not windswept and has extensive tracts of tundra, and thus caribou habitat, in its central highlands.

Caribou were abundant in Newfoundland in the late 1800s, numbering by some accounts as many as 200,000, but began to decline mysteriously towards the end of the century. The population fell steadily through the early 1900s until, by about 1925, it was nearly extinct, numbering perhaps 200 animals. The decline of caribou was a complete mystery, all the more so because it occurred as wolves, which are very effective predators of caribou, were exterminated from the island.[39]

The Newfoundland caribou recovered to about 3,000 animals by mid-century but were still only an echo of their former glory and retained very shaky prospects. It was not until the mid-'50s that the Provincial authorities undertook to seriously examine what had happened. Biologists began to examine the essential facts and before long it was clear that the problem had something to do with the production of calves (or, more precisely, the survival of calves); some 70 per cent of female caribou giving birth were found later to be missing their calves.[40] Newborn young of the deer family, including caribou, are pitifully fragile and exposed to a host of dangers including the basic challenge of getting on their feet. Survival of young moose, deer and the like is thus often very poor but the shocking loss of young caribou was extraordinary and was clearly the agent that was holding the population back. But what was killing the caribou calves? The tundra is a barren, essential place; there is simply not much out there to cause such a thing.

Biologists explored the blustery tundra and, sure enough, found a great many dead caribou calves. Most of the carcasses were consumed or decayed but many were still fresh and there was something very odd: nearly all the calves had peculiar pus-filled abscesses on the upper throat. Cultures were taken of the wounds and the infectious agent was identified as *Pasturella multocida*, a common and potentially lethal bacterium. The bacteria were, presumably, killing the caribou, but where was it coming from?

It was at this juncture, in the early '60s, that the young A.T. (Tony) Bergerud arrived on the scene. Bergerud, an intense and brilliant man, went on to become the world's leading authority on caribou (and a leading proponent of the idea that predation *did* matter) but was, in the summer of 1964, focused on the great enigma of the barrens of Newfoundland.

Bergerud spent days wandering the tundra examining caribou carcasses and made one intriguing observation: nearly all the dead calves were in lowlands. This was odd because the lowlands were protected from the barren's bitter winds and young caribou were known to be susceptible to chill-induced pneumonia. Calves should therefore have been dying more frequently on the uplands, not the lowlands.

Bergerud knew that lynx also preferred the swampy lowlands. One day, an associate found a freshly killed carcass upon which the abscess on the neck had not yet developed and there were four evenly spaced puncture wounds, two above and two below the throat. Bergerud rushed home to his skull collection (every biologist has one), retrieved a lynx skull, and matched the upper and lower canine teeth exactly with the puncture wounds on the calf's neck. He then retrieved a freshly killed lynx carcass from a trapper, took a swab of mucus from mouth, had the material cultured, and (as he was certain he would) turned up *Pasturella multocida*.[41]

Lynx were killing the caribou.

This conclusion might seem to have been less than fully inspired – the researchers, after all, must have known that lynx were the only predator present. The mysterious abscesses had, however, in every previous case, grown over and obscured the tooth puncture marks. Also, the habits of lynx were well understood even then, and lynx had never previously been reported to prey on caribou. The Newfoundland biologists had, nevertheless, been overlooking what was, in retrospect, fairly obvious.

Bergerud is possessed of a very inquisitive mind and next turned his thoughts to the history of events. Newfoundland is far enough removed from the mainland of North America to have a very limited diversity of its fauna; many species of mammal that are common on the mainland never migrated across to the island. Newfoundland in fact has only fourteen indigenous mammals.

Snowshoe hare, a common, and at times profusely abundant animal on the mainland, were one that had never made it across. (Hares, incidentally are close relatives of rabbits but have larger ears and do not burrow.) Someone introduced the snowshoe hare to the island in 1860, hoping the hare would establish as a source of food for the often impoverished islanders. The hare "took' and grew steadily in numbers until the early 1900s when the population, having depleted its food supply, first "crashed." The hare population remained depressed for some time before recovering to fix thereafter into the ten-year-cycle so familiar for snowshoe hare on the mainland.[42]

Bergerud astutely observed that the fate of Newfoundland's caribou had followed precisely the track of the introduced hare; the caribou "crashed" at the turn of the century just after the hare did and did not begin its modest recovery until the hare population did some years later. Given what he had seen on the barrens this could not, in Bergerud's mind, have been a coincidence. Records suggested that lynx were rare on the island prior to the introduction of hares (some observers question whether they were present at all). Lynx are superbly adapted to hunt snowshoe hare upon which, on the mainland, they rely almost totally for food. Lynx therefore grew with the introduced hare population and began to spread over the island, appearing in places (St. John's, for instance) where they had never been seen before. (A bill to exterminate the lynx was proposed in the Newfoundland Assembly in 1904.) The lynx, Bergerud proposed, "switched," as did Algonquin's wolves, to an alternate prey (caribou) when the hare population crashed with disastrous consequences for the alternative (Bergerud himself first coined the phrase "prey switching"). To compound the caribou's plight, the crash of the hare population coincided more or less with the building of the Newfoundland railroad which transected the caribou's main migration route. Hunters took this advantage to shoot excessive numbers of caribou and the caribou, faced with pressure from two new predators, simply couldn't cope.[43]

Case of the great Newfoundland caribou crisis closed.

There was (or is, for the saga continues) another actor in this drama. Newfoundland, it turns out, has its own native rabbit or hare: a unique, and now endangered, race of the Arctic hare of Canada's far north. Arctic hares are a larger and rather striking relative of the snowshoe hare, having black-tipped ears atop a brilliant snowy white coat that (unlike the snowshoe hare which molts to brown in summer) remains white all year round. Arctic hare are, like caribou, adapted to live out on the tundra. With the introduction of snowshoe hare, the Newfoundland Arctic hare went into a long decline and are now confined to several out-islands where the snowshoe hare never established, and a few sites in deep interior barrens so bleak as to be eschewed even by the resourceful snowshoes. The population that once occupied the entire island is thought now to number less than 1,000 animals.[44]

It was long assumed that the Arctic hare were simply displaced by the more aggressive snowshoe hare; out-competed for food and cover. Bergerud, however, demonstrated experimentally that Arctic hare can coexist with snowshoes (although at lesser numbers) in the absence of lynx but that if the snowshoe hare bring lynx with them the Arctic hare are doomed. Lynx, "switch," as they do with caribou, to Arctic hare when the snowshoe hare decline (as they inevitably will) and decimate the former.

Why, though, are the Arctic hare so vulnerable to lynx? Why can snowshoe hare easily sustain lynx predation while Arctic hare essentially throw in the towel? Here, Bergerud had another flash of insight. He studied geographical distribution of lynx, snowshoe and Arctic hare, and noticed an almost perfect alignment of the northern limit of lynx and southern limit of Arctic hare; the range of Arctic hare extended south everywhere exactly to the point they contacted lynx, but no further. There were a few places where Arctic and snowshoe hare coexisted, but none with lynx. Bergerud also noticed that the range of Arctic hare tended to match areas of the far north where snow, although present nearly year-round, never gets very deep and is compacted by tundra winds to an almost cement-like surface. Bergerud examined specimens of Arctic hare up close and noticed that their feet were conspicuously smaller than those of their southern cousins. Snowshoe hare (and lynx) have huge feet, the toes of which they can splay out and achieve very low pressure or "foot loading" on snow. The two are superbly adapted to run on snow, essentially snowshoeing (and thus the name) at very high speed. Bergerud deduced (and actually observed) that the larger and poorly-shod Arctic hare floundered helplessly in deep snow, easy prey for the lynx. This alone accounted for their range on the mainland and the rout they suffered on Newfoundland.[45]

Newfoundland's Arctic hare are trapped deep in what biologists call the "predator pit." They are trying continuously to recover their former abundance and range, to climb, that is, out of the pit, but every time they make gains, the snowshoe hare population collapses, lynx turn their attention to the Arctic hare, and bury them deeper.

I had an experience with Newfoundland's Arctic hare that drove home very personally the severity of their plight. In the fall of 1976, I found myself at the University of New Brunswick preparing to start post-graduate studies. My supervisor, Dan Keppie of the Faculty of Forestry, and I were looking for a good research project and, having read of the fascinating work with hare and lynx on Newfoundland, decided to go up and take a look. We were particularly interested in the Arctic hare and made the tortuous journey up the remote northeast coast of Newfoundland, our objective being the Grey Islands – one of the very few places that the Arctic hare had found refuge. A colourful, tobacco-chewing fisherman whose thick dialect made anything but the most basic conversation impossible (talk about a

Distinct Society!) took us across the ten-mile channel to the main island. (On the trip out, incidentally, I spotted a dolphin and, announcing so with great delight, was astonished to see the fellow produce a shotgun and take a shot at the animal – dolphins apparently being hard on the fishery. He, fortunately, missed and I kept quiet the remainder of the trip.)

The island was at the same time both breathtakingly beautiful and depressingly bleak. The latter impression was underscored by a walk we made one mournful day through an abandoned fishing village. The scene was positively spooky as we skulked under grey skies between rows of ghostly wood frame houses, the wind howling dismally, as in the cliché, through creaky shutters. I was immensely relieved to leave the eerie scene. The villagers, we learned later, had been relocated forcefully by a Government intent on bringing the island peoples to better schools and hospitals. This revelation did nothing to ease the melancholy that I carried for some time.

We, at any rate, searched high and low for Arctic hare for four days, finally catching a fleeting glimpse of the black-tipped ears of one as it raced away in panic at seeing, no doubt, its first humans. One hare in four days! (and this, you recall, on an island without lynx or snowshoe hare). At that point a wild storm brewed up and our fisherman was delayed in retrieving us for three more days, a period during which we ate nothing but salt cod that had been stored on the island for just such an event (a dish, incidentally, that I have avoided ever since). We abandoned the project.

In spite of the compelling evidence he had gathered, Bergerud had skeptics, some biologists being just unable to accept that lynx could so decimate a caribou herd. Bergerud, determined to prove his case, reverted to a simple ploy: remove the predator and see what happens. He had trappers remove lynx from two caribou calving grounds for two years. Calf survival improved dramatically on both areas and Bergerud's case was irrefutable.[46]

Predation did matter, or at least it did to Newfoundland's caribou.

Biologists continued to struggle with the issue of whether predators limit prey populations but, by the time Tony Bergerud left Newfoundland, a new way of thinking was emerging. The works of Errington, Gullion and others were regarded with near reverence, but it was becoming clear that things were just not that simple: predators did, at least in some cases, have the upper hand.

The ploy that Bergerud used to finally prove his case had an almost charming simplicity: remove the predator and see what happens. If the prey increase to any extent then the predator, obviously, was suppressing the population. Others have used the same strategy. Swedish researchers, for example, removed foxes and martens from islands off the coast of Finland to see if Arctic hare (which are "circumpolar") increased (they did).[47]

Doug Pimlott did it too. In an astonishing act that would be unthinkable, certainly controversial, and possibly even litigious today, Pimlott and his team killed 106 wolves in the Park in late summer of 1964 and '65.[48] Their motive was not experimental as with Beregerud and the Swedes; they simply needed various tissues and organs from the wolves to examine such things as the ages of the animals from which mortality rates could be determined. They must have felt that, at the end of the study, the removal was justified. Doug Pimlott and his assistants had great respect for wolves and the highest of ethical standards and, in the thinking of the time, I suppose the end justified such means. It is a measure of how things have changed to point out that such a thing would be inconceivable today.

Actually, in a way the experiment was repeated when the predators (Park Rangers) stopped killing wolves themselves shortly thereafter. In this case the prey apparently did not increase, probably (as discussed earlier) because the wolves responded by producing fewer pups.

Old habits, though, die hard. Well into the '70s (and even today) wolves were viewed by many as mere pests, vermin to be eliminated at every opportunity. This attitude was, and still is, based on exaggerated fears of the effects of wolves on deer. The hard edge of this was revealed in a shocking incident that occurred right along Algonquin's busy southern thoroughfare, Highway 60, in December of 1970.

Perpetrators that were never caught poisoned a pack of wolves just before Christmas that year. The pack had killed a deer on Eucalia Lake only a short distance from Highway 60. Someone laced the carcass, to which the wolves were returning to feed, with strychnine.[49] The incident might have gone unnoticed had not the Park naturalist gone to check out a report that a golden eagle (rare in Algonquin) had been seen feeding on the deer – and found five dead wolves. Of course, many other scavengers, large and small, visited the site and the known toll, along with the wolves, included four ravens, a fox, and probably the eagle.

Perhaps the worst part of the whole sordid event was that it was almost certainly Park staff who did the killing. As we have seen, Park Rangers had been systematically killing wolves (defended as "predator control") for many years before the Pimlott study, in vain and misguided efforts to protect the Park's deer. Some of the older fellows probably had some strychnine hidden away for just such an opportunity and – well, old habits die hard. Whoever was responsible must have spent some anxious moments, as Park naturalists made certain that the incident was widely publicized.

Bergrud's other ploy, the trick with the lynx skull, was employed to great advantage in the Park in the fall of 1986. We were looking at the time for brook trout spawning beds in lakes. Now, brook trout are normally just that, *brook* trout, and they spawn in well-oxygenated water in fast flowing riffles in brooks and creeks. The great majority of Algonquin's trout, however, are found in lakes. The water in lakes doesn't, of course, flow and brook trout have to find spots along shore where cold groundwater springs, so-called "seepages," percolate up through gravel. Such sites are very important to the trout and sensitive to disturbance, and so we had been documenting their location in the Park for some time.

An assistant was checking one such site on a brilliant Indian Summer's day when his eye caught a flash of pink on shore – brook trout acquire a deep scarlet colour when spawning in fall (more to come on trout in a later chapter). Investigating, he found a half-dozen dead trout scattered among the cedars. Thinking this at first the work of a poacher, we were about to contact Park Wardens when reason prevailed – poachers would not have left the fish at the scene!

I had just been reading Bergerud's work and, after reducing the possibilities to a short list (mink, bear, otter, raccoon and possibly an osprey or heron), grabbed several skulls and went to the scene. There were, sure enough, puncture wounds on the backs of the trout and the sharp canine teeth of the otter skull matched the wounds perfectly. Otter were killing our brook trout.

Word of the incident got out and several people, fishermen all, argued forcefully, with a curious echo of our wolf-abusing past, that we trap out the otter to protect the trout. I must admit to having had some sympathy with their views, but it's a Park, and Parks are places where otters should be free, if it is their pleasure, to kill trout, and wolves to kill deer.

# 5 | And Hares ... and Bears

*Who would not rather be numbered with the lion than the vulture?*[1]
— Robert Blumenshine, 1992.

We hold the grand drama of predation in awe, but we would perhaps be less reverential of our predators if we knew how often they participate in a much less reputable activity – the highly unsavoury act of *scavenging*.

Scavenging, the act of consuming a carcass that one has not killed has, as they say, a "bad rep." There is something fundamentally abhorrent about picking over rotting remains and we give thought to the scavenger – vultures, hyenas and the like – with only a shudder. Scavengers receive nothing but our contempt, but we would perhaps be kinder if we appreciated how wide their ranks really are.

We tend to be blinded by delusions of the state of decorum in nature. Take, for example, the noble lion, King of Beasts. Lions were once thought the *beau sabreur* of the African plain, dominating their realm with great integrity and courage. Lions are now known to scavenge extensively – almost as much, in fact, as the freeloading jackal. The bald eagle, symbol of liberty and namesake of lethal fighter jets, subsists mainly on dead fish.[2] Scavenging, it turns out, is a very common habit of many of the predators we hold in such high esteem.

It is certainly not practised by an exclusive club. No sooner is something down in our north woods than it is set upon by a host of ravenous opportunists. In winter, practically everything that moves, from blue jays to wolves, joins in the fun. Ravens (the northern cousins of crows) are first on the scene and use one of their eighteen different vocalizations,[3] a sharp "yelp," to summon their kin. These and other scavenging birds are so voracious and quick to find wolf kills that they may take a significant share of the meat from its rightful owners. Conversely, their activity and raucous calling will attract wolves to a carcass from miles (but ravens, in turn, respond to wolves howling in summer as to a dinner bell!).

It took a most determined and resourceful investigator to show us just how important scavenging is in Algonquin Park. Mike Wilton is a dedicated biologist who did marvellous work in the Park, mostly with moose, in the '80s and '90s. Wilton loved to challenge convention and was convinced that researchers were underestimating how very much predators like wolves and bears depended upon scavenged food. He made the most ingenious use of time-lapse photography to prove his point. Wilton put road-killed deer and moose out in a secluded spot near the Wildlife Station and rigged a camera set to shoot at various time intervals to a 12-volt battery. (He had, ironically, to cover the battery and wiring deep with large rocks to discourage a persistent bear from "scavenging" them!). The carcasses were no sooner put out than they were descended upon by a swarm of scavengers, mainly ravens, foxes, vultures, wolves, an otter, and the (presumably very hungry) bear. One 480 kg moose was almost entirely cleaned up in a few days in early June.[4]

Wilton made the very good point that previous studies had ignored scavenging at night, something his round-the-clock photos did not. The shadowy scenes he captured of bears and wolves staring malevolently at the camera, the flash reflected in their eyes, while tearing at the disembodied moose and deer are chilling, yet at once compelling. Viewing the ghostly images, one gets the uncomfortable sense of intruding on some primordial ritual not intended for human eyes, something akin, almost, to invading the confessional.

Wilton had shown that individual wolves and bears could scavenge with Teutonic efficiency, but there was a more important question: how much could scavenging provide food for predators like wolves and thereby reduce or "buffer" predation? The question was important because scavenging, if really significant, had the potential to relieve pressure on that part of a prey population that really "mattered" – the living part.

There is certainly no lack of stuff out there to scavenge. A study in the interior of Alaska, for example, showed that fully half the moose that died in winter were killed by something (disease, malnutrition, etc.) other than predators (and thus became "scavengable" material).[5] But how much of this perfectly nutritious flash-frozen meat is actually used by wolves? The question was vitally important, and several researchers would soon look at scavenging in this larger context. A key study among these was in Algonquin Park.

Graham Forbes of the University of Waterloo came upon the Algonquin scene in the late '80s to, among other things, answer that very question. Forbes, working under the direction of John Theberge at the University of Waterloo, took up the torch from Pimlott and Voigt to look once again at the ever-changing food habits of Algonquin's wolves.

The Park's wolves are smallish and, although they can and do kill moose, they are not notably adept at it. Forbes, and Theberge, determined by (sure enough) collecting scats, that wolves were nevertheless eating a considerable amount of moose in winter. Fully half of one winter's consumption, for example, was of moose meat. They guessed that much of this was scavenging but could not, of course, prove so by looking at scats alone. By tracking radio-collared wolves they were, however, able to locate and examine a considerable number of freshly dead moose carcasses. Applying a series of rather macabre criteria worked out in Alaska, they were then able to distinguish between carcasses the wolves had killed and those they had just scavenged.

The distinction is based largely on the fact that a moose dying from disease or malnutrition seeks solitude and shelter, usually in evergreens, and lies down in a collapsed posture with legs folded under the body to face death. These animals are often found frozen dead with the body upright. A moose killed by wolves, on the other hand, is almost always found in the open and lying on its side with all four legs extended. This basic distinction aside, there are more obvious differences such as the lack of blood and evidence of conflict, broken branches for example, at the scene. Scavenged moose are usually found frozen and consequently only certain easy-to-get portions, such as the upper shoulder, are eaten. The liver, a great delicacy, is unreachable. Finally, wolves perform a sort of ceremonial wallowing in the snow after a kill. If one didn't know better, it is as if they are trying to cleanse and absolve themselves of the murderous act. (Wolves, of course, don't feel "guilt" or any emotion as we do.) True to their enigmatic nature, they will not wallow before eating a scavenged carcass. One is tempted to suppose that they sense they are blameless (but again "blame" or "guilt" is a completely foreign notion to anyone but we humans.)

Forbes and Theberge, at any rate, determined that fully 83 per cent of moose eaten by Algonquin's wolves was scavenged,[6] placing them, for the three winters of his study at least, in the dubious company of African lions.

As if the knowledge that many of our revered predators are ignominious filchers was not hard enough, we now have to deal with the fact that we ourselves may be scavengers by heredity. The case for this was made very neatly in a recent article in *Scientific American* by Robert Blumenshine, an anthropologist, and his colleague John Cavallo. The two point out that that one reason scavenging by humans has received so little notice is that anthropologists are, ironically, only human and perhaps too quick to project current ways of living into the past (or, put another way, to reject our forerunners' seamier side).

We do tend to hold a romantic view of our ancestors: Man the Hunter, the noble savage taking the Saber-Toothed Tiger at spear point. There is, to be sure, ample reason to believe that our more recent predecessors were serious hunters. The Cro-Magnon "cavemen" of the recent Ice Ages left ample evidence of their hunting prowess, including details of big game kills in the marvellous cave art at Lascaux and elsewhere in southern Europe. If, however, one goes back much further in time, the case for hunting becomes murky. Just as our impression of lions and eagles has changed, so has that of our distant ancestors. The first proto-humans, the famous "Lucy" and her kin of 3 to 4 million years vintage, and even the first beings recognizable as "us," the earliest *Homo*, are now thought quite likely to have been essentially scavengers.

Lucy (who, incidentally, it turns out was probably a male!)[7] was of a type of small, graceful pre-human called Australopithecines. Much of the early impression of Lucy and her kin as hunters comes from the common discovery of Australopithecine bones in association with bones of "game," especially small gazelle-like animals. The pre-humans, it was naturally assumed, killed the deer. Recently, however, a pioneering "taphonomist" (one who studies events at kill sites) advanced the less inspiring and rather grim hypothesis that hominid and deer bones ended up in the same place because that is simply where predators (leopards) dropped them from their favourite feeding trees! Leopards, which then and now "cache" their food in trees, apparently took Lucy and gazelles with equal facility.[8]

There are other lines of evidence. Large herbivores are hard to bring down – they have ways of fighting back – yet the Australopithicenes, and even our more recent ancestors of one or two million years ago, were slight of build, almost gracile, not robust killers. There is also no evidence of true weapons among very early pre-humans; tools for scraping hides and crushing bones, yes, but not real weapons. Pre-humans also had the same dentition (types of teeth) as we have – the flat grinding molars and small canines of omnivores, not the shearing teeth of true carnivores. Finally, bones of large prey found with proto-men often show both tooth marks of carnivores and fractures from hammerstones, evidence that they were killed and scavenged by different agents. The evidence for Man the Scavenger is compelling.[9]

An argument against the scavenging hypothesis is that much of the meat available came from animals that had died of malnutrition or disease and would be of poor quality and particularly low in fat. This apparently can bring on a strange form of starvation known as "rabbit fever," from its common appearance in the 1800s in backwoodsmen who fed exclusively on lean game. Blumenshine and Cavallo point out, however, that primitive man was able to use heavy tools to crush skulls and large bones and get at brains and bone marrow, both rich in fat. (An ability we share only with hyenas, which have an exceptionally powerful bite.) Brains and marrow, being encased in bone, are protected from blowflies and the like and the last to putrefy (a process that, even in tropical Africa, can take as long as 48 hours). The evidence, in fact, is that primitive man largely scavenged the bony piles of defleshed carcasses. It seems that bone marrow and brains sustained our ancestors (or those, at least, with strong stomachs) and the cooperative foraging skills that must have been developed to secure such delicacies contributed greatly to our evolution as social animals.[10]

Blumenshine and Cavallo further point out that scavenging, while dangerous in itself, is probably no more risky than predation. There are, after all, hazards in the act of killing and, as we have seen, a warm carcass soon attracts a host of uninvited visitors, some of which can be very nasty. The visitors will often, in their words, "ignore the dead in order to pursue the living."[11]

Cavallo, determined to make the case for scavenging, spent many hours studying the behaviour of leopards in Serengeti National Park in Tanzania (leopards, or something very like today's leopards, are believed to be the cat most likely to have shared habitat with Lucy). Cavallo found that leopards and other large cats often leave their kills unattended for long periods, as long as twelve hours, ample time for the intelligent pre-humans to sneak in for a snack. This was very risky, but the risk in woodlands, where they could escape into trees, was probably tolerable. On open plains, with lions about, it probably was not.

The case for man the scavenger was further advanced with detailed observation of contemporary primitive peoples in sub-Saharan Africa who, to the astonishment (and horror) of the researchers, showed an avid interest in scavenging.[12]

It seems that we humans cannot, alas, place ourselves above the vulture, hyena and raven.

Scavenging aside, no one disputes that wolves kill most of the moose meat they eat. Under certain conditions, a very hard snow crust for example, wolves can take moose down almost at will.[13] The deep snow and strong crusts that develop in late winter will sometimes permit wolves to race, almost skate, across the surface. Moose break through the crust, foundering helplessly in the deep snow and wolves kill the hapless beasts with relative ease. In such conditions, even Algonquin's smallish wolves can tackle the biggest, strongest bull moose.

It is, in fact, widely believed that, in any but ideal conditions, Algonquin's wolves are rather inept moose killers. Every now and then, though, they surprise us. In the early summer of 2000, two Park Rangers were canoeing on Raven Creek in the northwest of the Park when, rounding a corner, they were astonished to see a bull moose on shore being set upon by four wolves. Two of the wolves were clamped onto the moose, one at each end, swaying to and fro in a grotesque sashay from the hapless animal's efforts to dislodge them. The smallish, dog-like animals seemed almost silly assailing the huge bull, like the tiny songbirds one occasionally sees chasing hawks. Silly, perhaps, but deadly in effect, for the moose was done for. Dazed, and with entrails hanging from its torn belly, the stricken animal staggered into the bush. The two uncommitted wolves had retreated immediately at the Rangers' approach but the two that had been "engaged" hung around for a minute or two, and then departed on the blood trail of the moose, apparently in no great hurry to return to the task. The Rangers reported later that (apart from being torn apart!) the ill-fated moose had appeared robust and healthy – a bull approaching his prime.[14]

Predation matters

Actually, it may at times be bears that are the more effective moose killers. Bears are omnivores, but are much more inclined to eating plants than meat. Black bears, for example, will subsist on nothing but grass for weeks. For this reason, and the fact that black bears have virtually never been seen to kill adult moose, biologists had for years disregarded them as serious predators.

There was, however, an odd thing about moose that defied explanation and suggested that some unknown predator was operating. Examination of the reproductive organs of female moose showed that considerably more moose calves were being born than were subsequently seen with their mothers after emerging from the hiding period after birth. This could, of course, be attributed to many things; moose, for example, are known to be highly vulnerable to pneumonia after birth. After Bergerud's revelation with caribou, however, the suspicion grew that some predator might be killing very young, or "neonatal," moose calves as well.

In the early 1980s, a fisherman happened, by incredible chance, to observe and photograph a black bear kill a moose calf alongside a trout creek near Espanola in Northern Ontario. The bear, surprisingly perhaps, had some difficulty with the task, having the calf slip from its grasp several times before prevailing in the unequal match. And there were other instances. For example, Mike Wilton (of the time-lapse photo work) unearthed the remarkable observation of a fire towerman near Missanabie, Ontario, from journals dating to the 1940s. The towerman had observed a cow moose with three calves pass daily on a trip to a low marshy area followed by a large black bear. In the towerman's words, "The trio of calves was noted to reduce to two and later to one with no knowledge of the cause. The same sequence of events continued with the cow, one calf and the bear, until it was eventually noted that the bear was no longer evident. A walk one day along the direction that the animals regularly followed toward the lake, disclosed the dead bear, badly mangled by the moose, from all the tracks that were evident."[15] In this case, at least, there had been a settling of scores.

Spurred by the accumulating evidence, biologists began to seriously examine the question of bears killing moose in the early 1980s. The ever-curious Wilton carried the Algonquin side. In the spring of 1983, Wilton collected stomachs of bears from hunters in the rough "bush" country that surrounds Algonquin Park on all sides. He subsequently examined, in meticulous detail, the contents of all of 296 bear stomachs, looking for remains of moose or deer. Before describing Wilton's results, the reader should be appraised of just what an appalling task this really was. The contents of predator's stomachs have (believe me) a singularly vile character, being a disgusting, shapeless grey mass of hair, bone and half-digested flesh that smells something like a blend of very dirty socks and rotten cabbage. To sift through such rot for hours on end requires a dedication to the task that verges on madness. The challenge is deepened by the fact that the tripe one receives may have been sitting neglected in the back of a pickup truck for days before being delivered, thus acquiring an additional piquancy.

I speak of this from experience, having personally examined the contents of hundreds of lynx stomachs (recall my experience with Cat Scratch Fever). Bear stomachs, though, are at least *interesting*. Lynx exist, for all practical purposes, on one thing and one thing only:

snowshoe hare. This was well established before our study but we had to be sure, and I (being the junior man on the research team) was assigned the task of confirming that our animals adhered to the tedious norm of lynx gustatory habits. I thus spent days in a stupefied torpor scouring over rancid lynx entrails in the vain hope of finding a little evidence of panache in the lynx lifestyle. I recall the great thrill I felt upon occasionally finding a grouse feather, mouse skull or squirrel toe amidst the great grey clots of hare parts.

Wilton, on the other hand, found what he was looking for. Almost 30 per cent of his bear stomachs contained remains of moose or deer.[16] Much of this was of adult animals, most probably winter kills that had been scavenged, but nearly half of the hair was of newborn fawns and calves, many of which must have been killed. (Wilton mostly found remains of deer but, had he been able to sample in the Park, which, of course, he could not, he would undoubtedly have found more moose.) Subsequent work in Alaska and elsewhere confirmed what Wilton's data suggested: black bears were killing a great many moose calves – enough, perhaps, to contain the growth of moose populations. The most convincing evidence of actual control came from another Bergerud-style "removal" experiment.

In the early 1980s, biologists in Saskatchewan had become seriously concerned about a moose population that, despite their best efforts, remained chronically depressed. Aware of the growing suspicions about bears, researchers "removed" 38 bears from two study areas of about 100 square kilometres each in 1983 and '84 (this was actually a lot of bears, the generally accepted number of black bears in similar forest type being about one per 4 to 5 square kilometres). Response of the moose population was dramatic; by fall there were more than twice as many calves per cow moose in the removal areas than two zones where the bears were not removed.[17] Clearly, black bears had been killing a great many moose calves in Saskatchewan.

As a result of all this work, the view that biologists held of black bears changed overnight from that of big but harmless oafs to potentially serious predators. But not everyone was convinced. Some biologists argued that the killing was "incidental"[18] (something along the lines of the "doomed surplus"). As with many of the issues in this most inexact of sciences, the debate was unresolved and continues to this day.

A book about Algonquin Park cannot avoid discussion of the most dramatic form of predation: predation on man. There were two fatal bear attacks in Algonquin Park in the past century, both of which, curiously, had multiple fatalities. In the first, three boys were killed in the spring of 1978 while trout fishing in Lone Creek on the Park's east side. The second incident happened in October of 1991 on Bates Island in Lake Opeongo. In the latter case, a middle-aged couple were killed at their campsite but not before one or both put up a heroic fight (long bruises were observed on the bear and a broken oar was found at the scene). Both people were killed by a blow to the head. Their remains were not found for several days, the bear having dragged the bodies some 400 feet from the scene of attack.[19]

The earlier attack was particularly tragic because three young people were killed. What exactly happened can never, as with the Bates Island attack, be known for sure because there were no survivors, but investigators pieced together the probable scenario. The three boys (ages 12, 15 and 16) were fishing and had had some luck because the first to be killed had several trout in his pocket. He had wandered some 200 metres from his mates when the bear seized him

*The gravesite of Captain John Dennison, age 82, killed in a bout with a bear in 1881. The grave, surrounded by a cedar rail fence, is being inspected by Harry Tuvi (left) and Nick Martin. The photo was taken in November 1978.* Algonquin Park Museum Archives #4505 – MNR.

and dragged him into the bush. The other two boys, it appears, went to his aid and apparently surprised the bear, which turned and killed them also. The bodies were not found until the next day; a 275-pound adult male bear was standing nearby, guarding them. The bear was shot immediately and autopsies on the boys and bear left little doubt that it was the bear at the scene that had done the killing.[20]

The bear was perfectly normal – a healthy adult male, in good condition – and not rabid (rabies is extremely rare, but not unknown in bears). This, as we shall see, is typically the case; people automatically assume something is "wrong" with killer bears, but they are almost always perfectly healthy large males in the prime of life. In further search of an explanation, it was postulated that the scent of the trout in the boy's pocket had triggered the attack, but this too is unlikely. Ocam's Razor, a time-honoured principle of scientific inquiry, states that the right solution to a problem is usually the most simple and straightforward one and the hard truth is that the bear that killed the boys was probably just hungry.

Bear attacks continue in the Park and may seem to be becoming more frequent. A forestry worker was attacked but not injured in 1999 (his dog distracted the bear) and a lady escaped serious injury by leaping off a small cliff into Big Trout Lake in June of 2002. Managing these bear problems poses a real risk to Park staff.

Such events are tragic but must be put in perspective because attacks by black bears are still exceedingly rare. There were fewer than 50 fatal attacks on humans by black bears in the last century,[21] fewer than one every two years on the entire continent (black bears range across essentially all of Canada and nearly all of the forested areas of the continental U.S., including even Florida). One is much more likely to be struck by lightning in Algonquin Park.

Peter Smith, a veteran of the Wildlife Station of whom we will hear much more in a later chapter, relates a story that is more typical of how wild bears behave. Peter was assisting with a study of pine marten (a type of woodland weasel) and the work involved having to bait live traps with a mixture of raspberry jam and cod liver oil. After some days of this, Peter acquired a distinct odour. A particular black bear caught the enticing scent and began to rendezvous with Peter every day at the same time and place along the trapline. At first, the bear, behaving as bears normally do, took off, but as the week progressed the animal became more and more familiar, eventually approaching to within a few metres. Peter actually reached for his hunting knife at one point (today he'd have pepper spray). Ever the dedicated fieldman, Peter carried on, but was sorely happy when the scientist in charge of the project called it off. The evolution from fear to familiarity to outright boldness is typical of how bears become acclimated to people, and eventually become chronic problems that have to be "put down." As we shall see, however, it is not this sort of bear that one needs really fear.

Rare or not, bears do attack, and bear attacks have been a source of morbid fascination for years and the subject of several very good books.[22] In part because bear attacks are so infrequent, the question *why?* has been asked time and time again. If black bears are not particularly inclined to attack people, what causes the attacks that do occur? Several researchers, most notably Stephen Herrero of the University of Calgary, have spent years probing every detail of recorded attacks looking for a common thread – the presence, for example, of food or pets at the scene – in hopes of discovering the "trigger," something that might perhaps be managed or eliminated to save lives. They have, one and all, been frustrated.

There is, nevertheless, some fascinating theory about bear attacks that relates to a fundamental difference between the evolution and behaviour of black and grizzly bears. The greatest threat to bears is, ironically, other bears. Adult males, both black and grizzly, regularly kill young of their own kind and particularly very young bears that have split from mom and struck out on their own.[23] This seemingly senseless behaviour may actually have a role in population control since infanticide in black bears may to be more common when food supplies are low.[24] (It is important to point out that the adult bears that do this killing are not in any sense "conscious" of a strategy. They are simply doing something that their genes have programmed them to do, because the bears that do this are more likely to survive (i.e. they obtain food from the young bear) and pass on their genes. Evolution works in this cold, completely efficient way; there is no "plan" by the bears.

Bears, at any rate, have enemies, including their own kind and (as we have seen) wolves, and, of course, man.

Black bears, however, enjoy one great advantage. The black bear is an animal of the deep forest interior and, when threatened, almost always has the option of running up a tree. Mother bears with cubs use this dodge routinely. Black bears do not therefore need to be highly aggressive; they can escape from danger. Grizzlies, on the other hand, live in open country – the mountain meadows of our far west and tundra of the north. When a grizzly bear is threatened, it has no choice but to stand and fight and anyone caught between a mother grizzly and her cubs is in grave danger. Biologists believe that living in such fundamentally different habitats has shaped the character of the two bears. Black bears, although smaller than grizzlies, are perfectly capable of killing people but rarely ever do (they don't have to). Grizzlies, out in the open and exposed, have become unpredictable killers and are far more likely to attack people than their laid-back eastern cousins.[25] (That said, there are roughly equal numbers of black and grizzly attacks simply because black bears are far more numerous and widespread. In fact, polar bears are by far the most dangerous of North American bears, regularly attacking without provocation. Fortunately, very few people share the polar bear's domain.)

*Climbing to safety.*

There is an odd twist to this concept. Herrero believes that most grizzly attacks are defensive – people blunder into the bear's space and the animal, feeling threatened, attacks. Black bears almost always retreat when encountered in the wild. However, on the rare occasion that a black bear does attack a person, it appears to be plainly and simply predation; the bear is looking for something to eat. There are several lines of evidence to support this. Very nearly all black bear attacks are by large males – big, aggressive animals that seem, simply, to be hunting. Secondly, black bears almost always *eat* their victims, something you would expect a predator to do (grizzlies do not usually eat their victims and, in fact, stop their attacks when the victim stops moving). Finally, there are additional cues from the way black bears behave. Black bears employ an entire repertoire of warning signals when they are threatened, things like jaw-snapping, false charges, and a harsh cough-like HUFF call. These warnings are a signal that you are too close – "YOU ARE IN MY PERSONAL SPACE – BACK OFF" – but seem to be strictly defensive and rarely precede a bona fide attack. Black bears that do attack approach quietly and deliberately, giving all the appearance of stalking you, which, in fact, they are.[26] As is often the case with people, it is the strong silent type you need worry about.

The message from all this is that when one encounters a black bear, and it does not immediately flee and continues to approach you, the best way to deter an attack is to act aggressively yourself; give the bear the impression that he faces a tough opponent. The bear is "assessing" whether to start or continue an attack. Screaming, throwing rocks or any type of extreme aggression is appropriate and may deter him.

*A large male bear. A close look reveals heavy wire on the stump to the left. The animal was a "nuisance" bear and was captured and relocated.*
Courtesy of Norm Quinn.

The really amazing thing about bear attacks, though, is how very few there are. There are well over 100,000 black and grizzly bears in North America and, with the millions of backpackers and canoeists using the wilderness, there must be millions of human-bear encounters every year. Bears are showing extraordinary restraint, not because they are particularly "noble," but because through time it has been the bears that usually come out the worse in encounters with people.

I received a graphic demonstration of this attitude one day in mid-summer of 1990. I was portaging a canoe through dense bush in the vicinity of Kingscote Lake at the extreme south end of the Park. The day was exceedingly hot and humid, one of those rare days in which Park acquires a near-tropical feel. I was staggering up a hill under the weight of the canoe, aware of nothing but that which my hunched posture brought to view and a sharp pain in my back. Unbeknownst to me, a bear was coming down the trail in the opposite direction; we were on a collision course. The bear must have been as torpid as I, for we did not make acquaintances until he had nearly stepped under the front of the canoe. Shocked out of my stupor, I yelled, dropped the canoe and, forgetting the standard wisdom about "standing one's ground," turned and ran. Recovering my dignity after a goodly sprint, I turned just in time to see the bear, which was a big one and probably a male, scampering over the hill, by all appearances as terrified as I was.

I wish I could relate a more heroic bear story but that, I'm afraid, is typical of my many encounters with bears. I generally see only their rear ends as they bolt away in panic. Still, I am vigilant when in bear country and advise you to be the same.

Just as the views that biologists held of predation have changed over time, so have those of the general public. People who live and work in rural areas are gradually becoming more tolerant of the presence of predators. The change, though, is occurring slowly and still has a considerable distance to go.

Paul Errington made a poignant appeal for predator conservation at the end of his magnificent work *Of Predation and Life*, calling predators "a manifestation of life's wholeness."[27] Errington, though, understood the deep-rooted barriers to people's acceptance of predators noting that it seems "almost a general characteristic of humankind to see depravity in predacious acts and to impute human ethics to wild predators."[28] Errington, you'll recall, was no ivory tower academic. He was a country boy, a hunter and trapper, the sort of fellow who, in the inimitable words of a former prominent U.S. politician, would "know enough to pour piss out of a boot."[29] Errington, no doubt, knew that, but he also knew how people in rural middle America thought and, even with his enlightened outlook, knew that there were absolute limits to the conviviality of people and predators. He pointed out soberly that economically vulnerable rural people and predators with expensive appetites do not usually get along all that well.

Predators can do great harm and there are situations that justify their control. For example, Algonquin's only truly threatened species, the wood turtle, is in jeopardy across its entire range in part because certain predators which prey heavily on turtles (especially raccoons and skunks) are at high numbers due to land use changes. (One could argue that that is our "fault," we having changed the landscape, but the problem remains in any case.)

Emotion and the hard issues aside, it is encouraging to see, as I have seen around Algonquin the past eighteen years, that some of the old attitudes are dying. Perhaps the day is not far off that wolves, bears and even the lowly coyote will be viewed as the great assets they really are. It will probably be some time yet, though, that we see wolves accepted in everyone's back yard.

# 6 | Of Time and Trout

*Where the pools are bright and deep*
*Where the gray trout lies asleep*
*Up the river and o'er the lea*
*That's the way for Billy and me.*[1]
  – James Hogg (Scottish poet, 1770-1835), 1838.

Eleven thousand years ago Algonquin Park lay under a mile of ice. The great Wisconsin Ice Sheet had advanced some 100,000 years earlier, as ice had three times before, covering all of North America as far south as Missouri. About 10,000 years ago the climate warmed and the ice began, with "glacial" slowness, to retreat north, revealing the Park's landscape of pulverized ancient hills.

For a time, Algonquin looked much like Baffin Island does today. The climate at the edge of the retreating ice was very cold, supporting an Arctic flora and fauna. There can be no question that then, if not much later, caribou, and probably muskoxen and Arctic wolves roamed the stark scene. Pollen cores show that a sub-Arctic vegetation of black spruce and tundra herbs existed at the time in the vicinity of Found Lake in the south-centre of the Park.[2]

The ice gradually melted away to the north and the Park warmed and its forests slowly changed, becoming more and more like forests to the south. In time, Algonquin acquired the mantle of Maple, Beech and Pines it has today. This is how we generally view the Park: a grand panorama of unbroken, cathedral-tall forests. In retreating, though, the glacier revealed and helped form what is perhaps an even greater asset: the Park's crystal-clear trout lakes.

The bitter cold glacial meltwater at first pooled into immense seas and Algonquin Park was revealed, temporarily, as a desolate, icy island. The receding waters coalesced into the pockets and grooves provided by the rugged rock shield and debris (the "till") dropped by the glacier and, in time, took the shape of the hundreds of lakes and creeks draining the Algonquin highlands today.

The frigid meltwaters were pioneered by coldwater fish like trout and whitefish. Others – the bass and musky of warmer climes – were excluded and, for the most part, remain so today. The trout found their way into nearly every corner of the Park, settling into the crystal clear headwaters as each lake formed and eventually becoming isolated, free to evolve on their own for 10,000 years. This isolation in space and time nurtured what has become a natural resource of immense value today.

There are over 1,400 lakes in Algonquin Park and over 300 contain natural lake or brook trout (by that I mean trout populations that reproduce themselves and are not supported by stocking). Prior to settlement, cold, clear trout lakes like these were common in highlands throughout much of northeastern North America. But no more. All around the Park the original native trout stocks are gone or greatly reduced. Pollution, overfishing and the introduction of "exotics" have wiped out all but a few scattered remnants of the original trout stocks of the Northeast. Algonquin, in fact, contains the only significant complex of more or less intact natural brook trout lakes remaining in the developed parts of Canada and the United States. This is an absolutely priceless natural heritage.

But that is not the half of it. Fish, you see, are cold, wet and slimy. Warmer, cuddlier things, things closer to us like moose and deer, get more attention and it is natural for people to think first of the Park's wildlife and forests. But the Park's forests, as magnificent as they are, are a long way from "natural." Uncontrolled logging in the 1800s and early 1900s fundamentally altered the face of Algonquin's forests and it will be generations before a truly "natural" forest recovers, even in wilderness areas (although the great majority of visitors will never know the difference).

Most of the Park's lakes, on the other hand, are untouched, virginal, in almost primordial condition. Anyone going to the considerable difficulty of travelling into Algonquin's deep Interior will be paddling over waters that, it is only a slight exaggeration to say, are as pristine as anywhere on earth. Find yourself alone someday in the middle of one of Algonquin's hundreds of small headwater lakes and you will be immersed in a scene and "system" that has not changed in any meaningful way since Native People were the only callers some 300 years ago. The plants, plankton, insects, fish and water itself – in short, the entire ecosystem – is whole.

The lakes, then, are the thing, but trout are the attraction. Algonquin's lakes are a magnet to fishermen. The Park attracts some 50,000 fishermen per year, almost all in about six weeks after ice-out in April when the fish are actively feeding in shallow water. Casting inshore for brook trout or trolling in big water for lake trout, anglers enjoy success rates that are typically three or four times higher than

*Tom Thomson, with a day's catch from Westward, 1912. In the background, upper left, are old mill buildings.* Algonquin Park Museum Archives #188 – O. Addison.

outside the Park. Despite the imposing numbers of fishermen, crowding is (generally speaking) not a problem. The Park is so large and there is so much water that fishermen can enjoy an essentially solitary experience. It is not unusual to have sole use of a small lake for days.

This is a good thing, for the Park's trout fisheries, as healthy as they are, are highly vulnerable to overfishing. The Park sits on hard granite and the soils that have developed from these rocks are poor in nutrients. Because of this the amount of energy that can be transferred up the food chain, from algae to plankton to minnow to trout, is limited. The lakes are, in short, unproductive. Trout populations in small lakes are shockingly sparse; the fishery of a typical brook trout lake of about 100 acres will be supported by as few as 60 fish in the spawning stock. It is something of a wonder that such small and isolated populations have persisted through the millennia. One would suppose that some sort of catastrophe would have struck at least once in all that time.

Isolation, and the time to adapt, may actually be the key to survival. God, it is often said, made trout last after he had practised on the other fish.[3] His handiwork, though, may reached an apogee in Algonquin Park where so many trout stocks have evolved in splendid isolation.

Let me explain. Isolation is how things change, how things evolve. Populations must be isolated to develop into new species, otherwise the flow back and forth of genes through interbreeding homogenizes everything. Fish, being confined to water, often find themselves isolated and change ever so slightly from their nearby kin.[4] Each of Algonquin's dozens of brook trout lakes have gone a little ways down this path.

Biologists used to think that evolution occurs very, very slowly, in imperceptible increments. We know now that evolution is like a low but bumpy hill with a great many sudden jolts on the slow road up. The hill itself is unimpressive but the bumps can be sudden and sharp. Evolution's bumps and jolts have actually been observed in nature, no more dramatically than in marvellous work with finches on the Galapagos Islands in the 1980s. A husband and wife team, Peter and Rosemary Grant, lionized in the best-selling *Beak of the Finch*, recorded very slight change in such things as beak length of the finches (the very same birds that sparked Darwin's revolution) as a result of sharp climatic change such as droughts. The Grants were able to detect this change, incredibly, in as little as a single generation. Each and every one of Algonquin's trout populations have been isolated for more than 3,000 generations (brook trout mature to spawn at two to three years of age).

Having 1,400 lakes in your Park can actually be quite a problem because people expect them all to be named. Aware of this, and having the bureaucrat's mind for order, a government official decided in the early seventies to tidy the books – that is, to give labels to thousands of hitherto nameless lakes in the north country. The job in Algonquin Park fell to two naturalists who took great pleasure in bestowing the scientific names of many of the Park's lesser creatures (birds mostly) on dozens of small, remote trout lakes. One can thus find the likes of *Dendroica* (warbler) and *Asio* (owl) lakes scattered across the Park's canoe map. One day, in a mischievous mood, they decided to name a lake BlueSea beside a lake already named Devil, the two lakes being separated by a short portage. The idea, of course, was that one really *could* be "between the Devil and the deep Blue Sea." Tragically for humourists everywhere, the proposed lists had to be vetted in Toronto and one cheerless but sharp-eyed official caught and rebuffed the ruse.

Things, at any rate, can change quickly and it is likely that Algonquin's brook trout differ from lake to lake in subtle, shadowy ways. But do these differences really matter?

For the answer, researchers first turned to an odd little fish called the stickleback.

Sticklebacks, of which there are several species, are intriguing little fish, well-known to aquarists for their charming habit of making tiny tubular nests from sticks and seeming to sit on the eggs like a mother hen. They are also the darling of an obsessive group of geneticists because the tiny fish bristle with a battery of bony spines, plates and rays – things that can be easily counted or measured. The first evidence that fish had been evolving or beginning to "split" into new species since the glaciers receded came from a study of sticklebacks in southern British Columbia.

Before things went all high-tech, geneticists used such counts to reflect the underlying variation of genes (everything, of course, being the product of a gene). The researchers in B.C. found that groups of three-spine sticklebacks that had been in the same body of water, but nevertheless isolated from each other since the glacier left, had developed tiny differences in their mouthparts. The differences were almost immeasurable but made a major difference in the fish's dietary habits.[5] A stickleback, it seems, is not a stickleback, three spines or not.

Sticklebacks are one thing, but fisheries biologists get really turned on by things you can catch and eat, and the next example of this genetic "fine tuning" came from brook trout. In 1991, a researcher in Rhode Island proved that wild brook trout in a hatchery setting could somehow distinguish between water from their home stream and the hatchery groundwater.[6] This was one of the first indications that fish inherit powerful abilities to orient to their home waters. Through the late '80s and into the '90s, evidence accumulated that fish do indeed have an astonishingly sophisticated innate "awareness" of their surroundings.

The clincher, though, had actually come from work done earlier with another trout-like or "salmonid" fish – Pacific salmon. In 1988, Donald McIsaac and Tom Quinn (no relation of mine) of the Washington State Department of Fisheries published results of a study of Chinook salmon from the Columbia River that rocked the close-knit community of fisheries biologists, and made headlines in science columns around the world (quite something, actually, for a fish, which have trouble pushing the bears, wolves, tigers and their ilk off the front pages).

Pacific salmon spawn in gravelly rubble in streams off the major river system of the Northwest. The life cycles of the several species of salmon differ somewhat but all follow a basic pattern. The fry develop for a time in their home stream but then move downriver and eventually out to sea where they grow to maturity and return to spawn at (generally) three years of age. Biologists had known for some time that salmon had an amazing ability to "home" to their birthsite but McIsaac and Quinn revealed this ability to have an almost surreal dimension.

One race of Chinook known as "upriver brights" were known to spawn at a certain stretch of a tributary of the Columbia known as

Hanford Reach. The two researchers intercepted upriver brights (which, as their name implies, are uniquely coloured) on their way to spawn at Hanford Reach at a fish hatchery 370 kilometres downstream. They spawned the fish and raised the eggs in the hatchery, releasing the fry the following spring (the young fish had thus never seen Hanford Reach, their ancestral spawning area). Three years later, right on cue, these "naive" upriver brights (which had been tagged) returned and, as one might expect, spawned at the hatchery but, incredibly, 10 per cent of the salmon kept going the 370 kilometres to Hanford Reach![7]

How, possibly, could this have happened? What mystical force drove the salmon, which had no prior experience of Hanford Reach, past dozens of suitable sites to find the exact spot on which their ancestors had spawned for thousands of years? Not believing in the occult, McIssac and Quinn knew that there must be some tangible, physical explanation for what they had seen. But what? What process or agent could possibly account for fish finding an ancestral site they had never been exposed to from thousands of miles out at sea? The answer has to lie somewhere in the salmon's genes for those genes are the only thing connecting the fish with their forebears (and Hanford Reach).

There is really nothing mysterious about the basic way that genes work. Genes produce chemical compounds called proteins and enzymes (a type of protein actually) that are the messengers and building blocks of our entire beings. McIsaac and Quinn speculated that a protein "lock" might exist in the scent organ of the fish that matched a unique chemical messenger from Hanford Reach.[8] This, though, is really just speculation. We may never understand the amazing migratory abilities of Pacific Salmon (which in a way is a good thing; it's nice to have some of nature's mysteries remain mysterious).

The implications, though, are enormous. Pacific Salmon may not be one homogeneous unit but rather hundreds of sub-groups, each biologically programmed to seek out its own birthsite. These stocks may be irreplaceable for if any one is lost, its home creek, used for thousands of years, may become barren, as all other stocks bypass it in search of their own.

The importance of the salmon study goes way beyond salmon and even beyond fish. Fisheries researchers are actually leading biologists to appreciate how important this sort of "fine tuning" can be to fish, and to wildlife management and biodiversity preservation.

Take, for example, Algonquin's brook trout. Is it possible that each of the Park's dozens of brook trout stocks are adapted in some fine way to the character of the lake it has been confined to for some 10,000 years? We know that brook trout in a few very small lakes spawn in deep water, seeking out the dark holes where cold, oxygen-rich groundwater seeps up through gravel. Should these fragile trout stocks be lost can they be replaced? Would other fish find the hidden springs? Probably, but it is a question we would rather not put to the test.

But what, you might legitimately ask, is the problem? Algonquin's trout are secure. They are indeed, but there is (or was) a problem, a real threat that was developed, ironically, to help the fisheries.

Next to the alarming loss of certain marine food fisheries, the proper use of hatchery-reared fish is perhaps the biggest issue in fisheries management today. Fisheries biologists, aware of the sort of work done by McIsaac and Quinn, are coming increasingly to question the use of hatchery fish in natural settings like Algonquin Park. Concern over this has, in fact, become so great that biologists in some circles refer to the hatchery product, half jokingly, as the "Devil's Spawn." This, certainly, is a bit much, but there is a growing agreement that hatchery-reared fish do not have a place in wilderness.

The problem, of course, is that hatchery fish will interbreed with the native fish, inserting genes that may dilute the "wildness" we now know to be so important.

Algonquin's lakes are certainly not immune to stocking. Up until 1986, when the practice was suspended in the Park Interior, up to half of the trout lakes in the Park had been stocked at least once (something of "Coals to Newcastle" approach when you think of it). The problem, in the case of brook trout at least, is that the source of the stocked fish was a long-established strain borrowed from a hatchery in Pennsylvania. What sense does it make to stock brook trout coddled for generations in a secure hatchery environment in Industrial Middle America into Algonquin's pristine lakes? Sense or nonsense, well-intentioned biologists did so for decades.

In the early '90s, two geneticists, armed with all the tools of modern science, set about to determine what effect all this stocking may have had on Algonquin's wild trout. Roy Danzmann of the University of Guelph and Peter Ihssen,[9] a scientist with the Ontario Government, scoured the Park for several years, collecting brook trout in every corner they could reach (taking only "fry" or young fish to minimize any impact on the lakes). Their hope was that the blueprints that genes show us today would give testimony to nature's ancient accounts and, more recently, that of humans.

They need not have worried for their efforts were richly rewarded. Algonquin Park contains five main watersheds, wellsprings of all the fresh water that radiates out to the surrounding landscape. Danzmann and Ihssen found that brook trout from each of these watersheds were genetically distinct, reflecting the ancient paths that fish took to colonize the Park after the glaciers left.[10] More importantly, though, Ihssen subsequently found that trout from *each and every* lake were unique and could be distinguished through their genes. Time and isolation had wrought their subtle effects. Trout from lakes only a few hundred metres apart could separated by their genes. (A benefit of which will, incidentally, accrue to the enforcement of our fishing regulations – "but officer, I caught the fish on – – —lake.")

But the great value of the work came in the revelation of the legacy of indiscriminate stocking. This was possible because inbred hatchery stocks invariably show unique mutations that are very nearly perfect "fingerprints." The stocks of hatchery brook trout in historic use in Ontario (again, the main source of which comes from Pennsylvania) all have such identifiers and Danzmann and Ihssen could therefore tell which of Algonquin's lakes had purely wild brook trout and which showed evidence of interbreeding with stocked fish. They had sampled brook trout from 43 lakes, 25 of which had, at one time or other, been stocked. Incredibly, *all but one* of the stocked lakes showed the hatchery marker gene.[11] (Two of the eighteen "not stocked" lakes did also, probably because the old records for these two lakes were lost.) Stocked fish, the "devil's spawn," had infiltrated every corner of the Park.[12]

So, stocked fish have defiled Algonquin's virgin waters, but I have a feeling that the Park's trout, having survived 10,000 years of adversity, will weather this storm also. As noted earlier, artificial stocking (except for research purposes) is no longer allowed in the Interior of the Park. It should not be long before nature reasserts itself and our brook trout adjust to any damage that may have been done.

In truth, much of the concern over stocking may be exaggerated, and stocking does have a legitimate role to play in fisheries management. Hatcheries, for example, can be used to conserve rare fish stocks. Ontario has a very rare and strikingly beautiful fish known as the Aurora trout (a variety of brook trout). Aurora trout lack the blue "halos" that brook trout display on their flanks but have replaced them with a stunning band of bright crimson. Aurora trout existed in the wild in only three lakes in northeastern Ontario that fell victim to acid rain. The fish are alive today because of a far-sighted hatchery manager's decision to establish a broodstock in the hatchery before the wild fish died out. (The hatchery stock eventually became the source of two natural populations in lakes that had been rehabilitated.)[13] Some aquatic lakes are so far gone that the only way to maintain any fishery at all is by stocking. So fish stocking is not the great menace that many claim, but few would dispute that is has no place in real wilderness.

Ironically, much of what we know about fish stocking techniques comes from research in – you guessed it – Algonquin Park. Jim Fraser, a venerable fisheries biologist of the old school (meaning he was in the field more often than the office), conducted studies on stocking strategies in Algonquin for almost thirty years beginning in the early '60s. Fraser was an affable, down-to-earth fellow who charmed everyone he met with his earthy wit and simple, self-effacing manner. He worked out of the famous Harkness Laboratory of Fisheries Research where he raised, in summer, every one of his six children (which at times both amused and annoyed the Lab's corps of scientists).

Fraser studied different stocking techniques (such as changing the "delivery system" – truck or aircraft) in dozens of trials to see what produced the best "return" of stocked fish. One spring in the early '80s, he conducted one such experiment that, for no apparent reason, was a dismal failure; practically none of the stocked fish survived. Puzzled, Fraser visited the lake, which was very small – only about five acres – to have a look, and happened to notice an apparently well-fed loon

*The Ontario Fisheries Research Laboratory at Lake Opeongo in 1937. The woman sitting on the steps is Mrs. Harkness, wife of W. Harkness, who set up this first research station.* Algonquin Park Museum Archives #3442 – Dave Wainman.

*The field camp of the Ontario Fisheries Research Laboratory at Costello Lake in 1937. Dr. A.G. Huntsman, holding a fish can, is seated on the running board with Mrs. Huntsman to his left. W.A. Kennedy is perched on the front fender. D.L. Robb leans against the rear fender while J.A. Arnason relaxes with a cigarette. V.E.F. Solman is at the wheel of this first fish laboratory truck. The others are not known. Note the printing on the side panel of the vehicle.* Algonquin Park Museum Archives #6231 – Don Smith.

paddling contentedly across the surface. Led to the obvious hypothesis, he tied plastic tags to the next batch of several hundred trout and stocked the lake again. Allowing a few days for the experiment to develop, Fraser returned to the lake and, with the appropriate permits of course, shot (or, euphemistically, "collected") the loon. Examination of the loon's stomach tuned up dozens of coded plastic tags![14] Loons, Fraser concluded, are capable of consuming prodigious quantities of trout and spoiling some expensive stocking programs.

Word of the study got out and prompted cries to "control" the loons from some circles. But as with the otter feeding on trout, and wolves feeding on deer, it is nature's way (and the right way) in Algonquin Park, and so no more loons were collected. (And, yes, Fraser would probably not be granted his loon-collecting permit today – even in the interest of science.)

Frustrated and wanting to improve the survival of his stocked fish, Fraser tried an ingenious experiment. With a truly vaudevillian flair, he contrived of a plastic model loon rigged to swim along a circular track in a hatchery raceway full of young trout. The idea was that the constant harrassment by the fake loon would train the young trout to avoid the real thing in the wild. After two years of this training, Fraser, confident of success, stocked his learned charges in a small lake having several resident loons along with an equal number of untrained trout. Alas, subsequent netting showed that the trained trout actually fared worse than the native fish![15] Brook trout are apparently very slow learners.

Jim Fraser pioneered in another aspect of the biology of brook trout – their fascinating spawning habits.

Most of Algonquin's lakes are headwaters – small lakes with very minor inflow creeks many of which dry up in summer. As mentioned earlier, brook trout confined to these lakes can't move up creeks to spawn (their preferred option) but search out sites near shore where cold oxygen-rich springwater upwells through gravel – so-called "seepages." In late fall, just before the lakes ice up, female brook trout congregate over these seepages and excavate nests, or more properly "redds," in the gravel by digging or "cutting" vigorously with their tails. Males, meanwhile, hover about the perimeter of the area vying

*Brook trout digging nest.*

for attention from the busy redd-builders. The fish eventually pair off (often actually with two to four males crowding the one female) and spawn, each female laying 500 to 1,000 eggs, which she then covers to protect while incubating over winter. Competition for the few suitable seepage sites can be fierce.

There is ample evidence, in fact, that the shoreline springs are at a premium, perhaps critically limited on some lakes. We did extensive surveys for spawning sites in the late '80s, and I know of one lake for which the entire trout population is supported by a single disk-shaped basin of springwater about the size of a bushel basket. Every fall, the entire spawning stock – dozens of fish – crowds over this spot, which is a mere pinprick along the shore, competing fiercely for space. Fraser, in fact, found that the desperate trout build nest upon nest, the eggs in lower tiers often suffocating.[16] More recent work showed that brook trout are able to "feel" spots along shore where the upwelling is strongest, and confirmed that there is not enough area of "seepage" to accommodate all the fish in most lakes. These inshore springs are among the most sensitive and critical fish habitats anywhere. Their rarity may explain why natural brook trout lakes have all but disappeared in Southern Ontario outside of Algonquin Park – where cottage development on lakes is extensive.

Finding these rare and vital sites was hard work – long, tedious hours paddling blustery shorelines in the late November cold. The effort, though, took me into some of the most remote and beautiful spots in the Park. I recall in particular the day we found a site on a very small lake in the Western Uplands, perhaps the most inaccessible part of the Park. We had been searching lakes in the area for days – enduring long paddles and portages in the bitter chill of late fall – but to no avail, the fish weren't cooperating. Late in the afternoon of a long day's search, we (I was working with two students) rounded a corner into a small bay of the last lake on our list and hit the jackpot: a spectacular site with dozens of brightly coloured trout milling about the shallows. Cold and exhausted, we lingered at the scene for some time savouring the stunning beauty of the brilliant crimson fish swimming through aquamarine water over the yellow and jade-green gravel bottom they had cleared. It occurred to me that we were quite possibly viewing something that no one had ever seen before. It is extremely unlikely that any tourists had ever been up there in mid-November, and even Native People might have found the spot a bit out-of-the-way. We left on this inspiring notion and with a deep sense of accomplishment that I have rarely known.

But back to fish stocking and genes. The trouble with all this is that fish which may have really important underlying genetic differences don't usually look any different. This makes the job of convincing the skeptical fishermen all the more difficult. He sees no difference and the hatchery fish end up in the same frying pan – so where's the problem?

Researchers believe that the superficial visible traits of fish, such as colour, are under relatively simple genetic control – sometimes only one gene.[17] On the other hand, things that may be important to fish managers, such as growth, are more likely to be controlled by complex groups of genes – so-called "polygenes." Therefore, fish from different lakes can *look* very similar but have hidden differences that are important to survival.

Algonquin's mysterious "silver" trout illustrate the point superbly. There are two lakes in the Park (which for obvious reasons will remain nameless) that contain rare mutant strains of a sort of "albino" or colourless trout. The fish from one lake in particular appear in every detail but colour to be lake trout, but are a pure, bright silver, nurturing the belief of local people that they are a type of salmon (which causes endless headaches for enforcement officers, because the catch limit for salmon is more liberal). Ihssen studied the fish in the late '80s and determined that they are, sure enough, lake trout but, genetically at least, rather ordinary lake trout.[18] The one difference that is expressed so vividly, colour, results from a single "mutant" gene hidden among thousands.[19]

Genes aside, one cannot write about trout without writing about colour, for trout, along with the peacock, represent perhaps the apotheosis of nature's proclivity for kitsch. A biologist from California has written of the beautiful little Golden Trout of the Sierra Nevadas in almost reverential terms: "Never had I seen such intense blends of greens, carmines, yellows, and black, yet it wasn't garish, nothing clashed, and the total impression was exquisite beauty beyond words."[20] Algonquin's brook trout are of the same genre, being composed of a stunning blend of crimson, olive, copper, jet black, white and sky blue. The result is striking at any time but intensifies to an almost riotous effect in fall, when the fish spawn. The dominant hue then is a deep scarlet-red which suffuses much of the surface of the fish.

And therein lies one of Algonquin's great natural allegories, for at the same time as the Park's trout are turning red so, in a much more heralded event, is her dominant tree, the sugar maple. The two events have fascinating parallels.

The stunning transformation of our hardwood forests in fall is a recurring delight, a display acclaimed by countless poets and artists, but few have written of the simultaneous and equally dramatic transformation of trout. This exquisite blush of trout and trees is superficially similar, but there are deeper, more substantial similarities as well. In both cases the phenomenon, as sublime as it is, plays a fundamental role in survival.

Trees do not actually "change" to yellow or red in fall. What happens is that as the temperature cools, leaves stop producing chlorophyll, the green pigment that drives photosynthesis. The remaining chlorophyll then breaks down and disappears, revealing yellow, orange and red pigments that had, until then, been concealed. The yellow and orange pigments are carotenoids, a class of pigments that include beta carotene (which our bodies turn into vitamin A). Carotenoids are stable in leaves, present from early summer, in species like aspen, but the brilliant reds of maples are caused by a different group of pigments – anthocyanins – that are formed just before the chlorophyll goes.

The anthocyanins, also responsible for the reds and blues of fruits like grapes, appear just in time to perform a crucial function. Leaves are an important store of nutrients, an asset that trees risk losing every fall. To prevent this, trees retrieve the nutrients from their leaves as fall progresses. One nutrient, phosphate, actually suppresses the formation of anthocyanins so its early departure from the leaf enhances the coloration. (Interestingly, the commonly observed reddening of leaves of crop plants, a so-called "hunger sign," is due to phosphate-deficient soil.) The onset of deep frost in late fall brings the salvage process to a halt and trees make a last ditch effort to retrieve their nutrients just prior to leaf fall. Anthocyanins play their role by protecting the dying leaves from frost, thereby extending for just a few days the vital retrieval.

An equally elegant mechanism occurs in our piscine counterpart, the brook trout. In this case, the operative pigments are the yellow-orange carotenoids (those same "carotenes" found in carrots and exhorted by health food advocates). Carotenoids, as a group, are becoming recognized as a sort of biochemical jack-of-all-trades, serving a broad spectrum of roles in plants and animals. For example, beta carotene, which cannot be made by fish or man and must be obtained in the diet, is split in half to form Vitamin A which, in turn, performs a whole array of vital functions from facilitating eyesight to controlling how cells develop. A particular carotenoid – astaxanthin – is stored in the muscle and liver of brook trout and it gives the flesh of trout, and relatives like salmon, an appealing pinkish cast.[21] The carotenoids (mainly astaxanthin) are taken from the liver and muscle in fall and delivered to the surfaces of the fish – skin and fins – to achieve an impressive display. This is most pronounced in the male trout and is thought, in the absence of any other reasonable explanation, to serve in the courtship of females. In females, however, the same compounds serve a higher function than mere ceremony. Female trout deliver their carotenes primarily to their developing eggs where they serve a critical role in survival – a role that is not fully understood but was first discerned in hatcheries.

Trout eggs are various shades of reddish-orange. Hatchery mangers have known for decades that eggs with deeper red-orange colour survive better, but the reason for this has remained elusive. An early suggestion that the colour played a role in fertilization by somehow attracting sperm to the egg has been debunked. Another theory would have it that the carotenoids afford protection to eggs from sunlight, but trout eggs are buried in sand and gravel after fertilization, casting doubt on that idea also.[22]

The explanation that is now generally accepted is that carotenoids somehow improve the delivery of oxygen to fish eggs – as they apparently do in mollusks (clams, etc.). Mollusks can survive in polluted oxygen-poor water if they have elevated levels of carotenoids. This idea is supported by the observation that the carotenoid content of eggs of several species of Pacific salmon is related to the velocity of water they swim in – those that spawn in slower water with lower oxygen have higher carotenoid levels. Also, fish spawning in the open ocean, where oxygen levels are high, have eggs that are low in carotenoids (although this may serve to make the eggs less visible to predators). Carotenoids may even help the eggs survive in the low-oxygen environment of the belly of female trout before spawning.[23]

Finally, there is an ironic connection between trout, carotenes and man. Much of the trout available in supermarkets is not actually wild but raised in fish farms. Such trout are fed commercial fish pellets and can develop a pale, unappealing flesh tone, a problem with marketability (wild trout obtain carotenoids from crustaceans, such as crayfish, in the natural diet). To rectify this, fish farmers often supplement the trout ration with carotenoids prior to market, thus giving the flesh a more natural pinkish hue. This seems an almost satirical turn of events; an "additive" used to create a cosmetic effect but inadvertently duplicating the natural, benevolent state of affairs.[24]

Brook trout from various lakes in the Park actually do look quite different in their fall glory, exhibiting varying patterns and intensity of crimson, pink and blue. These differences are due in part to diet but must also reflect underlying genetic variation, the real importance of which we may never understand.

The hordes of fishermen who descend on the Park in spring are after one thing – trout. But Algonquin offers another option for fishermen, an option available in the warmer months – bass. Smallmouth bass are not native to the Park but were deliberately introduced into about 80

*A good catch, even by the standards of the early days – Rock Lake, 1916.* Algonquin Park Museum Archives #2000 – Mrs. Swan.

*This photograph of a 20-pound Salmon Trout (Lake Trout) caught in Algonquin Park ultimately became a postcard.* Algonquin Park Museum Archives #6415 – George Garland.

lakes beginning as early as the late 1800s, the idea being to provide for some fishing in the summer (trout retreat to deeper, cooler water in summer and can be hard to catch). Such a thing would be unheard of today – we do not play such games with the Park's lakes – but it has to be admitted that the experiment worked, with thousands of campers enjoying fresh bass grilled in Algonquin every summer. Bass, furthermore, provided the subject for one of the neatest pieces of fisheries research ever done in the Park, or anywhere.

Bass are not really comfortable in the North country, being a "warmwater" fish better adjusted to more southerly climes. This was evidenced as early as the '50s when researchers noted sharp differences in the abundance of young bass in Lake Opeongo from one year to the next. Fishermen have known this for years – some years fishing is obviously way better than others. This is what biologists refer to as "year class strength," the year-to-year variation in production of young. In some species, like bass, year class strength varies greatly and biologists come to speak of year classes as "the great class of 19 – ," much as connoisseurs speak of wines.

The first inkling that year class strength of bass was determined by weather came in 1957, when W.J. Christie of the University of Toronto noted that the years in which small bass were abundant on Lake Opeongo were almost always preceded by a warm summer three years previously (bass grow rapidly and first reach catchable size at three years of age). It seemed that warm weather in the first summer of life gave the young bass a head start.[25] Data on the bass fishery of Opeongo were collected for decades and Christie's early intuition was confirmed by a sophisticated statistical analysis done in 1981. This work showed that water temperature in the first summer of life is, sure enough, important, but the severity of the following winter may also be critical.[26] Young bass, it seems, need to "bulk up" and put on a lot of fat to get them through the long winters when they are inactive and do not eat – in essence, hibernating. A short, cool summer does not give the fingerling bass enough time to feed and, if followed by a long winter, can be disastrous. The tiny bass starve by the millions. As such, Algonquin's bass are not unlike our deer, bear and even moose which need to put on substantial reserves of body fat to get through the long winter.

# 7 | Stress – Vulnerability Abounds

*To the man who is afraid, everything rustles.*[1]
— Sophocles, 5th Century B.C.

The next time you're in a job interview, get a call from a tax auditor or are called upon to do some public speaking (the universal dilemma), take note of what your body is doing. The racing heart, dry mouth and sweaty palms, familiar to all but the most serene of us, are part of what scientists refer to dispassionately as the "Standard Stress Model." Stress, and its detrimental effects on health, have become a trendy obsession in modern society. But few people realize that *exactly* the same process that so plagues us also troubles wildlife and can have serious consequences to the health and well-being of wild animals from frogs to moose. Stress, in fact, is an abiding concern of anyone working with wild animals and growing awareness of the potential effects of stress has led to some of the most fascinating thinking and theory around the enigma of wildlife cycles.

We tend to think of wild animals as being tough, way "fitter" than we are, and capable of prodigious endurance. This may be true in the strictly physical sense but, emotionally, wild animals can be extremely fragile and this vulnerability to stress can quickly erode their physical vigour. Many wild animals seem to be on an emotional hair-trigger and even mild trauma can send them into the potentially fatal spiral of the stress response.

Wild animals can suffer acute stress from the seemingly most innocuous of things and yet show no sign at all. An early demonstration of this came from work with deer in Upper New York State in the '70s. Aaron Moen, a professor at Cornell, was doing innovative research on the strategies deer use to survive winter just about the time that the snowmobile came into popular use. Professor Moen outfitted semi-tame deer with remote heart monitors and documented that the passage of snowmobiles, even at great distance, caused a marked increase in the heart rate of the outwardly placid animals.[2] This, as Moen pointed out, was no trivial thing for, as we have seen, deer are often near starvation in late winter and need to conserve every calorie. Snowmobilers, nevertheless, won the day and the cursed things are now everywhere (although not, of course in Algonquin Park where their use is prohibited).

My first noteworthy experience with stress in wildlife was in 1985 when we in the Park were involved in capturing and transferring marten to the state of Michigan. The purpose of the project was to repopulate a part of the state from which they had been exterminated. (Algonquin Park has a time-honoured history as a source for such restorations – but more on that later). Marten are a type of weasel, beautiful, sleek creatures about the size of a skunk, which come in a wonderful variety of shades on a basic brown theme from bright orange to almost black. Marten are highly efficient and frenetically busy predators searching incessantly for their preferred and nearly exclusive prey, a type of mouse called the Red-Backed Vole. While not really common anywhere, marten are nevertheless widespread across the North and have long been a mainstay of the fur industry.

We were live-trapping marten throughout the fall and briefly confining them in holding pens prior to veterinary inspection whereupon they were delivered to the grateful Michigan authorities. On one typically grey November day, I had retrieved a beautiful male specimen and, stopping to make a phone call on the way to the station, left the animal in its holding cage in the back of my truck. The call turned out to be prolonged and I had not noticed that a light but very cold rain had started. When I returned to the truck I saw that the animal had got wet – not soaked you understand, but slightly damp. Thinking nothing of it (after all, this was a wild animal that lives constantly exposed to extremes of weather) I covered the cage with a tarp and proceeded on my way. It was shortly after I arrived at the Wildlife Station that I noticed the marten in distress, shivering and emitting a strange sort of grating noise. I watched in helpless dismay as the animal deteriorated rapidly, and by morning it had curled up and died. An internal examination showed bloody scars along the surface of its intestines, telltale "lesions" that indicate death in wild animals from an acute syndrome that accompanies severe stress. The marten, its delicate emotional state being upset by capture and confinement, had been pushed over the edge by a minor physical challenge and died in very short order of, essentially, shock. It was an unkind demonstration of the fragility of wild things that I have not forgotten. (I should point out that the project was, on the whole, a great success and marten are now happily ensconced in Michigan.)

The process that felled the marten that day was simply a rather extreme example of the same process that bothers us in our day-to-day lives and has become perhaps the leading popular obsession of our time. Stress has become a universal and unremitting fixation of the media.[3] We are informed relentlessly that we are "stressed out" and in dire need of treatment to the body or psyche or both. Entire industries of physical and pharmaceutical therapy have developed to conquer the common foe.

The stress response evolved to deal with the occasional but severe dangers faced by our primitive ancestors. The problem today, as is well known, is that the same surge of adrenaline that prepares the body to tackle a saber-toothed cat, the so called "fight or flight" response, is triggered continuously by the trivialities of modern life – traffic, ringing phones and the daily dirge of depressing news. An occasional surge of stress is not harmful, but continuous exposure to adrenaline and other stress hormones wears the body and its defenses down and can result in debilitating emotional and physical disease. (The American Academy of Family Physicians estimates that two-thirds of visits to family doctors are prompted by stress-related ailments and the three best-selling prescription drugs are for high blood pressure, anxiety and ulcers.)[4] Animals, even relatively "lower" animals like frogs, have essentially the same array of glands, organs and chemicals that deliver the stress response and seem to be at least as vulnerable as we are.

The stress response is a complex thing involving the flow of hormones like adrenaline and cortisol, which give that familiar "sinking" feeling in the pit of the stomach.[5] But it was, in fact, from a simple animal, a frog, that I received another graphic lesson on the effects of stress on wild things.

In the late 1980s, biologists from around the world noticed that frog populations were declining.[6] The realization, almost overnight, that frogs were disappearing from places as far apart as the western U.S. and Australia sparked a great deal of concern and excitement in the media. The great fear was that some single and sinister global event was at the root of all or most of the declines. Maybe the frogs were warning us of an impending doomsday or "end of world" catastrophe. Since then declines have been observed from Panama to Poland and almost everywhere else that herpetologists (people who study amphibians and reptiles) are active. The crucial question of whether there is a common cause or not is hotly disputed.[7]

I became interested in the issue in part because parks like Algonquin are perfect places to examine such questions. If sensitive species are declining in the relatively pristine environments of large parks, then the case for some sinister, omnipresent "global" cause is easier to make.

Interested in studying the stress response in frogs, I had arranged to meet with a young graduate student at the station who was working with blood parasites, for a lesson in taking blood from the creatures (quite an art, actually – much more difficult than from even the most atrophied human vein). The student, a young lady from the University of Toronto, remarked at the outset on how much difficulty they were having in keeping the frogs alive in captivity; most became severely stressed within hours and had to be released. By way of proof she pointed to a large bullfrog, one of her captives, in an aquarium to her side. I looked closely and noticed that the skin on the frog's extremities, the toes and lips, was gone, and there were blotchy spots elsewhere where the skin was deteriorating. Fascinated, I watched for the next hour as the poor frog continued to, almost literally, disintegrate. The student was clearly distressed, but I must admit I was simply amazed for the erosion of the animal's extremities was progressing visibly. Under the stress of capture, the frog's immune system had shut down and it was being devoured by a rapacious army of bacteria that its white blood cells normally kept in check. The impression was of a sort of bizarre amphibian equivalent of the dreaded "flesh eating disease" (or more properly necrotizing fasciitis) of humans.

Most wild animals, in fact, harbour and are in constant battle with an array of potentially lethal bugs. These are mainly various forms of bacteria and in particular Salmonella, which, given an opportunity, will rapidly overwhelm a weakened immune defense. Capture and captivity seem to be particularly dangerous and zookeepers, upon receiving a new charge, use antibiotics as a matter of course to suppress aggressive infections. The phenomenon is known as capture stress, and capture stress is precisely what killed my marten on that that cold November day. Predisposed to a stress reaction by capture and pushed over the edge by the chill, the animal's defenses failed and bacteria in the gut, normally held in check, proliferated, eating away at the intestinal lining and causing a fatal hemorrhage. That particular response is so common in stressed wildlife that blood in the gut is one of the first thing vets look for when determining cause of death of an animal.

In the early '70s, John J. Christian, a biologist studying at the prestigious Albert Einstein Medical Center in Philadelphia, became fascinated with the implications of the stress response in animals. Christian hit upon what he thought might be a comprehensive theory to explain, once and for all, the mystery of wildlife population cycles. Biologists were searching for an all-embracing explanation for animal abundance – a sort of "theory of everything" of population biology. Christian was aware of the devastating effects stress can have on animals, and that social interactions, the relentless pressure of the "peck order," can be a major source of stress. He reasoned that any wild population might be limited by stress effects long before it reached the limits of the resources (food, cover, etc.) in its environment. He published a seminal paper "Social Subordination, Population Density, and Mammalian Evolution" that eventually became a pillar of wildlife biology and sparked a debate that continues to this day.[8]

Christian conducted a series of experiments with captive mice (voles, actually) to substantiate his ideas. He observed that as the numbers of voles increased, fighting, particularly among males, increased to intense levels, reproduction ceased and the population "crashed." The voles were simply too "stressed out" to function.[9] Subsequent work with the voles and other animals confirmed that stress hormones rise sharply with increasing numbers and set off the incapacitating physical decay of the stress syndrome.[10]

Christian's ideas caught on in some circles and grew into something of a cult, spawning some of the more draconian prophecies of our time. The innocuous machinations of mice in a lab came to reflect our own inevitable descent into war, starvation and disease as the human population increased uncontrollably. Time I suppose, will tell, although there is some very recent evidence that human population growth may finally be slowing down.[11]

Stress has many faces, and stress of an entirely different nature made itself dramatically evident in the Park in the winter of 1985. The agent here was our familiar friend, the moose.

For years, biologists in Michigan's Upper Peninsula had been in pursuit of a dream – to restore moose to the state. Moose had been exterminated from Michigan in the early 1900s but Ralph Bailey, a biologist working in the Marquette office of the Michigan Department of Natural Resources, was convinced that the habitat and public mood was suitable to attempt a reintroduction. Ralph lobbied for years and eventually convinced his superiors to accept the idea. Ontario was Michigan's closest neighbour with moose, and the delicate process of negotiating a "donation" began in the early 1980s. (Contrary to popular belief there was no "trade" of moose for wild turkeys, the reintroduction of which was already underway in Ontario and has proven a great success.) Almost inevitably, Algonquin was chosen as the source.

In February of both of the bitterly cold winters of 1985 and '87, the skies of Algonquin Park throbbed with the staccato thump of helicopters as a small army of biologists, rangers and veterinarians conducted

*A moose being transported during the Michigan Moose lift.*
Courtesy of Peter Smith.

one of the most ambitious wildlife management projects ever. Fifty-nine moose were sent across the border in a superbly organized affair that was a model of skill, innovation and international cooperation. The project was ultimately successful and moose are now happily ensconced in the forest and lake country of the Upper Peninsula. (Growth of the population has not, however, met the more optimistic expectations possibly, in part, because of brainworm from local deer.)

The operation was conducted with two helicopters – a fast, manoeuvrable pursuit chopper which located the animals, drove them onto a lake and tranquilized them with a powerful sedative from a dart gun, and a second, much larger "lift" helicopter that transported the huge animals out in an ingeniously designed sling. At the base camp, a corps of specialists then processed the animals for shipment to Michigan in a precisely choreographed concert of probes, measurement and tests. I was a part of the team assigned to the lift helicopter.

Pursuit of the moose was sometimes difficult and prolonged. Some of the animals seemed to sense immediately what was up and instinctively sought shelter in the coniferous cover along shore. In such cases, the pilot and shooter, concerned for the well-being of the moose, would call off the chase after a few minutes. On one occasion, however, the chase went on for too long, or perhaps the moose was not particularly fit, for we were to witness a strange, sad and normally fatal syndrome that often strikes wild animals under the stress of pursuit and capture.

The processing of this particular bull moose went normally until, nearing the finish of its veterinary tests, its hind legs began to quiver and spread apart in the "ataxia" or paralysis familiar from brainworm. This, however, was definitely not brainworm, the effects of which never come on that quickly. Veterinarians at the scene recognized instantly what was happening and injected the animal with selenium. Selenium promotes the production of vitamin E, which can prevent the problem, but it was too late; the paralysis of "capture myopathy"[12] had set in and the animal had to be destroyed.

Stress, it seems, as with exercise itself, and in this case compounded by the exertion of the chase, raises levels of a lactic acid in the blood. Lactic acid is a by-product of exertion that, at very high levels, can actually destroy the muscle tissue. When this happens damage to the muscle is generally irreparable.[13] People (and animals) that are fit can burn much more oxygen through prolonged exercise before acid levels build to harmful levels (marathoners, for example, have a resistance to the acidosis of exercise that is superhuman).[14] The extreme exertion of a long chase can predispose an animal to the problem but, amazingly, simply being confined can also bring on the fatal syndrome. The moose we selected for capture that day was, apparently, and as strange as it sounds, not in very good "shape." (Farmers, incidentally, are very familiar with the syndrome which frequently occurs in pigs and other animals being sent to slaughter and is known as "shipping disease.")

Intrigued, I asked one of the vets how animals like moose can bear up through prolonged and obviously stressful chases by wolves without myopathy setting in. He guessed that quite often they don't, that some animals – the less "fit" ones – succumb early to stress and wolves may find an easy meal. It is strange but probably true that many human athletes have more staying power than wild animals, even those we consider "athletic" like moose and deer. (Moose and deer do get a measure of revenge on wolves; a researcher found that arthritis of the joints is very common in wolves in the Park, presumably from the many long and hard miles on the chase.)[15]

Another wildlife recovery project became the occasion for me to study stress again, although the stress in this case was, as we shall see, entirely my own.

The subject this time was otters. The state of Missouri, it seems, once supported a healthy otter population, but the animals had all but disappeared from the state by mid-century – a result of habitat loss and over-trapping.[16] When Ontario began looking for a source of wild turkeys in the mid-'80s, Missouri was mentioned as having prime stock, the wildest turkeys around. Upon contact, Missouri was happy to help but not, in this case, for cash, but for wild river otters, of which Ontario is replete, in exchange. A deal was struck to exchange "x" number of otters for "y" number of turkeys, the host jurisdiction being responsible for capture and delivery of its side of the bargain.

The deal was, I suppose, a sound one on paper, but it soon became apparent that Ontario got the raw end for we quickly learned that otter are fiendishly hard to catch alive.

Otters are a fascinating creatures. Another member of the weasel family, the otter is famous for its playfulness, and scenes of families of otters cavorting along riverbanks are common fodder for the matinee wildlife show. Otters are something of an enigma; curious creatures, they have an odd habit of approaching right up to canoes with what some find a threatening guttural snorting. Yet, for all this bravado, they are (ask any trapper) very wary of traps or at least live traps.

Ontario found this out the hard way when we first attempted to discharge our part of the bargain with Missouri. Algonquin (naturally) was chosen as the site to coordinate the project, and a set of large pens was repaired at the Wildlife Station to hold the animals until enough were acquired to ship. Initially, we simply provided live traps to local trappers with the promise of a healthy payment upon delivery of an otter in good condition. It quickly became apparent, however, that the otters (although by no means scarce) were not cooperating. Trappers called repeatedly in frustration with stories of being thwarted in a variety of ways by the wily animals which seemed to have a phenomenal sense of just where not to step (the live traps functioned something like giant clams, closing on the target with pressure on a centrally located trigger). The first season went by with only one otter delivered to Missouri and, frustrated and at risk of losing serious face with our neighbours to the south, we searched through the winter of 1985 for a solution.

Someone had heard of a research project in Alberta in which otters were captured and radio-tagged for tracking studies. We contacted the Alberta authorities and they tuned us on to the only man in the world, apparently, who can capture otters live – Ted Code.

Ted was a biologist by training but worked part-time as a trapper. Like most professional trappers (and biologists for that matter) he was something of an eccentric, but was also an extraordinarily capable outdoorsman and naturalist with a deep-rooted understanding of the ways and wiles of our target. He agreed to help, a considerable sacrifice because it meant abandoning his work at home, and arrived in the Park in October of 1986. Ted reminded me of a sort of thin Grizzly Adams. He was unusually loquacious for a woodsman, ready to spin a yarn almost anytime. Within days of starting he had his first otter and, through ploys and devices known only to him, had approached our quota by mid-November.

That is where my troubles started, for we had now to direct our thoughts to how we were going to get the otters to Missouri. One day I happened to be at our research labs north of Toronto on an unrelated matter and I struck up a conversation on the otter project at coffee with one of the technicians. The fellow had a great deal of experience working with captive animals and, when appraised of what we were doing, acquired a solemn demeanor and warned me not, under any circumstance, to be caught in a confined space with an otter. The vicious brutes, he explained, "go right for your #&*#'s." Alarmed, and not knowing of any way of getting the otters out of their pens without going in after them (to that point we had just been passing food and water in) I pressed him for details but received only more ominous warnings focused on the male accessories.

Now, otters, despite their reputation for playfulness, are formidable creatures. Basically a 12-kilogram tube of muscle with razor-sharp teeth at one end, otters are accomplished predators. I had once witnessed an encounter between an otter and a German Shepherd dog and recalled feeling very sorry afterwards for the dog. I began to view our charges for the first time as a serious problem. How on earth were we going to catch a dozen or so of the vicious brutes in a confined space and pack them into small cages for the long trip to Missouri, while maintaining our, so to speak, integrity?

I mentioned the technician's warning to Ted upon returning to the Park. Ted shuffled his feet and stared at the ground and, uncharacteristically I recall, declined comment, ominously avoiding even eye contact.

Shortly thereafter the problem came to a head when I received a phone call from Bill and Mary, two biologists at our head office who were in charge of the program, informing me that the move to Missouri was imminent. (Bill and Mary — for reasons that will become obvious — are not their real names.) Mary, who was in charge of details, was a meticulous and exceptionally organized person, devoted to the job, and determined that the project would go off without a hitch. This complicated my problem further because, upon contacting the U.S. authorities, she was informed that there was a prodigious volume of paperwork (export permits, veterinary clearances, etc.) and inspections required at the border. Determined to succeed, she thus insisted that I leave the Park with the otters for the first leg of the long trip well before sunrise to give us ample time to clear customs at Detroit during business hours. We would therefore have to catch the animals in the dark!

My colleagues in the Park and I deliberated at length in some distress on the best way to enter into a ten-by-ten-metre pen, capture twelve lithe, powerful and purportedly dangerous wild animals in darkness and stuff them into small holding cages. An adage for meeting difficulty, popular with the earthy local folk who live around the Park, came repeatedly to mind. It was "like trying to stick a wet noodle up a wildcat's ass," and was rather fitting when you think of it. After ruminating for hours on a number of hopeless proposals we decided there was nothing for it but to fashion some sturdy hand nets, notify the local hospital and trust to luck.

Thus prepared and fortified with strong coffee (and I having taken the added but covert precaution of wearing an athletic cup) we arrived at the scene at 4 a.m. on a cold and wet November Monday.

Along with an assistant, I set up gas lanterns, grabbed our nets and, holding our breaths, entered the den with an eerie sense of intruding. There followed a prolonged and chaotic melee as the two of us struggled frantically with the disobliging otters in the semi darkness. I captured one robust specimen almost right off and forced it into a small rectangular holding cage only to find the door jammed open. Squatting over the opening I watched in horror as the animal escaped between my legs, his jaws only inches from the aforementioned target appendages. To my great relief, the animal passed on the opportunity.

Shortly after starting, the otters' water trough capsized and the water mixed with otter urine and dung scattered about the floor, forming a slippery paste that added greatly to our travails. I then received a rather nasty cut from a steel edge and we shortly found ourselves struggling in an obnoxious goo of otter dung-paste, coffee and blood (mine mostly), a concoction that I don't suppose has been brewed up before or since!

Despite these impediments, the operation proceeded without serious incident. The otters, although incredibly strong and hard to handle, were not, to our surprise and great relief, overtly aggressive – their fearsome reputation apparently a sham. One by one we wrangled them into the cages and loaded them at length onto my van. Before doing so each animal was given an injection of a powerful sedative to thwart the onset of the stress response (the only real connection, it must be admitted, this story has with the subject at hand). I proceeded down the road, exhausted and filthy, but a wiser man. (In retrospect we should perhaps have clued into the benign character of the animals from the fact that although they had been confined in close quarters for weeks, and certainly capable of doing each other harm, we never once saw them fighting.)

I arrived in Toronto to pick up "Bill" and "Mary" at about 7 a.m., already exhausted, yet faced with a two-day expedition to Missouri. Mary was in great anxiety over the state of our charges (which were sedated and sleeping soundly and in much better shape than I) and fretted all the way to Detroit about the great bureaucratic barriers we were about to confront, obsessively rearranging the mountain of permits she had so properly and painstakingly acquired. Her agitation was infectious and had us all in a great funk as we approached the border.

We arrived at the customs gate to face the proverbial intimidating agent with huge belly and sidearm to match. Asked what we had to declare, I replied with a straight face "twelve live otters," expecting to be ushered immediately into the halls of officialdom if not arrested outright. The fellow stared at me for a long moment and then replied dully, "SAY WHAT?" whereupon I repeated my statement of intent. He stared for a while longer, and then simply waved us through. And that was that! Mary, I suspected, was greatly deflated.

We drove on, passing through the monotonous landscape of cornfields that extend, it seems, without end through the level terrain of southern Michigan, Indiana and Illinois. We were determined, for the sake of our cargo, to do the long trip non-stop, but as the evening wore on we began to tire seriously. Before long, however, another more pressing problem developed. Otters, it turns out, have a particularly foul-smelling urine, something akin to a mix of skunk (another weasel) and ammonia, and our somnolent charges had been peeing copiously from the outset. Before we were out of Michigan the van stank atrociously. We nevertheless pressed gamely on through the night, Bill taking his turn at the wheel around midnight.

Now, Bill, as is common among wildlife biologists who often cultivate a rustic image, chewed tobacco, and had, since leaving Toronto, been spiting the exudate into a margarine tub he placed on the dashboard.

*The ever-inquisitive Otter.*

As the night wore on Bill, as a means to ward off fatigue, became more and more animated, regaling me with stories of certain of his exploits as a youth in the outdoors and elsewhere. Somewhere near Chicago, carried away with one such yarn and gesticulating wildly to make a point, he knocked over the margarine tub, spreading what seemed a quart of the filthy gunk all over the front of the van and, to my horror, me (I had the misfortune to be seated up front).

Completely disgusted, Mary and I used this event to force a halt. Exhausted, we checked into a cheap motel, Bill and I split a six-pack of watery beer (Budweiser, I believe) and we all got two to three hours of sleep (all that Bill, who actually had seniority, would allow, being determined to discount the incident with the spittoon and press on).

We started out again at dawn, turned south towards St. Louis, and drove all of the next day and night, arriving in Kirksville, Missouri, at daybreak. By now, the van was an appalling morass of otter pee, dung, phlegm and beer farts. (How Mary, who, in spite of her resolute bearing I suspected for a delicate soul, endured it, I'll never know.)

We were greeted by our counterparts from the Missouri Department of Natural Resources and an entourage of local politicians, media and locals (the project having developed considerable profile). Adroitly manoeuvering the eager dignitaries away from the interior of the van, we disembarked the otters and prepared to give our blessing and escape to the nearest bed and bath.

Now I generally like and admire Americans but find them at times insufferably forward. This group, being no exception (and after all Southerners), insisted, in spite of our haggard condition, that we accompany them without delay to the release site and post-release festivities. Having dutifully done so (but not before consuming, as is the local custom, a prodigious breakfast of biscuits and gravy), we attempted once more to disengage but were, to my amazement, thwarted again, this time with an invitation to tour the Missouri countryside. Bill could not diplomatically refuse and we spent the afternoon sightseeing in catatonic fatigue.

This accomplished, I assumed the ordeal was over and prepared to find a long-overdue bed but, to my absolute horror, we were all invited, with great fanfare and command, to a party! We were at that point experiencing the trance-like second wind of profound fatigue and beyond caring, and so spent the evening in a foggy reverie with our euphoric hosts consuming extraordinary quantities of pizza and watery beer (Schlitz this time, I believe). Released at last, I collapsed into the deepest sleep of my life. We awoke next day restored for the long trip home and learned in the art of transporting large weasels cross-continent. (The otters, incidentally, weathered the trip beautifully, forming the nucleus of a wild population in Missouri.)

Stress, as we have seen, can manifest itself in strange ways, but leaving my personal adventures, we can return to the weighty subject at hand and ask: Was Christian right? Can stress, in itself, limit wildlife populations?

Christian's ideas have received mixed reviews since he first put them forward. Results from laboratory mice, it seems, are not always applicable outside the lab (as is so often and sadly the case in medical research). External factors such as food and predators probably limit wildlife numbers more often than stress. That said, fresh support for Christian's old idea came very recently from a fascinating study in the Yukon.

The study involved lynx and snowshoe hare, the two players in the textbook example familiar to generations of biology students, of population cycles. Very briefly, snowshoe hare populations fluctuate in a remarkably consistent ten-year cycle and are followed very closely by their principal predator, the lynx. Biologists believe that the cycle is driven by food; as the hare population grows, their winter supply of woody browse is exhausted and the hare literally eat themselves out of house and home. The hare population then crashes, followed by the lynx. After an interlude of two to three years, the food supply, and then hare and thus lynx, recover and the populations grow, peak and crash again, and so on.[17]

The Yukon researchers measured stress hormones in snowshoe hare during a population decline and found, amazingly, that the sheer presence of lynx in great abundance caused the hare such stress that the population declined well *before* the food supply ran out. The hapless hare, being in a state of unremitting anxiety, developed extremely high levels of cortisol (the main stress hormone) and, as with Christian's voles, were too stressed out to breed.[18] The population of course then crashed, thence to recover in the classic fashion. It seems that just the enhanced *possibility* of being eaten by a lynx was enough to upset the hare's delicate emotional balance and imperil the entire population. The Yukon researchers, aware that their results were controversial and would be challenged, took great pains to rule out other factors and have, it seems, provided a later-day vindication of Christian's hypothesis.

Christian's ideas remain controversial but, right or wrong, his thinking has had a great and lasting effect on how biologists think of the animals they work with and handle day to day. There is perhaps not a better example of this than an elegant study done recently on African wild dogs in Tanzania.

Biologists have been darting animals with tranquilizing drugs and almost ritually measuring, poking, prodding and attaching radios to their quarry for many years. In fact, hardly any investigation of anything larger than a hummingbird is done now without the use of radios. There has, however, been surprisingly little research on the effects of all this usage on the subject animals. The African researchers, concerned about possible stressful effects of their handling and collaring on the wild dogs (a hyena-like animal that is one of Africa's most endangered mammals) took a most ingenious approach to the problem.

Cortisol, it turns out, is passed in tiny amounts in the feces of mammals under stress and can be measured in minute concentrations with highly sophisticated extraction techniques. The African researchers followed their study animals and a comparable "control" group of non-radioed wild dogs around for days collecting their droppings. (Biologists, you may have noticed, do some pretty strange things, many of which centre in one way or another around dung.) The droppings were then analyzed in the lab and it was determined that cortisol levels of the radioed and control groups were nearly identical – strong evidence that the process of radio collaring was not particularly stressful.[19] This is very likely the case in most radio-telemetry work, especially as the technology gets more sophisticated and the various collars and implants smaller and smaller. The Tanzanian study, though, is one of very few to take a hard look at the issue and a measure of the far-reaching impact of J.J. Christian's thinking.

# 8 | The Twig Eaters

*The animal needing something knows how much it needs, the man does not.*[1]
   – Democritus of Abdera, 5th to 4th Century B.C.

Wildlife, when you think of it, have a pretty tough go of things. We humans (those of us in the developed world at least) don't lack for nourishment, most of us being in a more or less continuous battle with our waistlines. Wild animals, on the other hand, are caught up in a never-ending search for food, a struggle for survival that they often as not lose.

There are exceptions. For example, the ruffed grouse, the familiar "partridge" of our north woods, lives in an apparent embarrassment of riches. Grouse are known to eat over 400 different types of plants in summer and fall[2] when the forest floor is an unlimited larder. In winter, grouse live entirely on the buds of trees, particularly those of aspen and birch, which are distributed in billions across the forest canopy in an apparent inexhaustible supply. Something other than food must limit grouse populations.

Species like grouse, however, are unusual; most wildlife rely on food supplies that are dangerously unpredictable. Deer are a prominent example and, as we have seen, can starve by the thousands in tough winters, but many of nature's "lesser" species suffer also.

Every year in late spring, just as the blackflies are getting intolerable, the Park is invaded by hundreds of thousands of diminutive but exquisitely beautiful birds known as Wood Warblers. The warblers, of which there are more than 30 species (20 nesting in Algonquin), are essentially unknown to most people and yet are extraordinarily common and come in an amazing array of brilliant colours with the accent on various hues of yellow, green and blue. Each species has its own distinct and melodious song which, in concert with dozens of other songbirds, produce the awe-inspiring "dawn chorus" of mid-May. Aldo Leopold, father of the modern science of wildlife management, wrote of a man who, when appraised of the fact that there were over 30 species of tiny songbirds of the same family in every colour of the rainbow within a few minutes' walk of his house, became convinced of the existence of God.[3]

The warblers, as successful as they are, have a serious weakness, for they are specialists and rely entirely on one foodstuff – insects. (Specialization is a dangerous strategy to follow in the natural world.) Anyone who visits the north country in mid-May becomes aware straight-away that there is no lack of insects, the blackflies alone coming in shocking hordes. The arrival of warblers in late spring coincides precisely with this glut of insects and they gorge furiously on nutrient-rich bugs to support the energy-demanding tasks of staking territories, nesting and egg-laying. (Unfortunately, warblers probably do not eat many blackflies; they generally "glean" insects and their eggs and larvae off foliage. Only a very few species of warbler catch flying insects.)

Occasionally, however, this feathered *fête* is tragically interrupted. Weather in the north is cruelly unpredictable and it is not uncommon in mid-May to see a sudden return to the sub-zero cold, ice and snow of Algonquin's recently departed winter. These cold snaps, if pro-

*Warbler in Maple with snowflakes in the background.*

longed, can spell doom to warblers, for insects become inactive for days on end, and the birds, their energy needs at a peak, starve *en masse*.[4] At variance with their cheery demeanor, warblers and other songbirds, in fact, face short and brutish lives. Most nestlings never see their first birthday.

Another familiar Park bird, the Great Blue Heron, regularly faces starvation. Herons, the common stork-like bird of the north, present an odd contrast of awkwardness and grace. The great birds are commonly seen in marshy shallows adeptly spearing frogs and fish with their long beaks and, once satisfied, leaving the scene with ponderous flight and a raucous *frawnk*. Any heron one sees feeding with such apparent aplomb has had to first face a stern test, for research has shown that many herons die from starvation shortly after leaving the nest. Young herons are apparently inept in the art of spearing frogs and those that don't learn quickly enough don't make it.[5]

We see then, that the provision of food is an ongoing concern for most wild animals. But it is more often the *quality* of food than the sheer amount that is problematic.

Most wild animals live on a shockingly poor diet. Spruce Grouse, for example, the northerly cousin of our ruffed grouse or "partridge," live exclusively on conifer needles (jack pine mainly) through the long

*A splendid male Spruce Grouse. This northern variety of grouse is much less common in the Park than the more familiar ruffed grouse or "partridge." Spruce Grouse are found in only a few spruce bogs on the south and east sides of Algonquin Park. Courtesy of Norm Quinn.*

winter months. The needles, or leaves, of coniferous trees are exceptionally poor in nutrition, having as little as four per cent protein,[6] a level that, theoretically, cannot sustain higher forms of life (domestic fowl, close relatives of the grouse, require at least 16 per cent protein to grow and lay).[7] How grouse can live on such an incredibly poor diet baffles scientists, and it is assumed that they compensate by eating prodigious quantities of needles, making up for the poor quality by stuffing themselves. Recent work in Alberta showed that another common denizen of our north woods, the snowshoe hare, are often forced to subsist through winter on forage that, theoretically, they should starve on.[8] Again, it appears that they just stuff themselves.

Some wild animals, spruce grouse being a good example, have an amazing ability to pick out individual plants among hundreds that have foliage with very slightly higher levels of nutrients ("palatability" in the jargon). Studies have shown, for example, that the pine trees that spruce grouse pick to feed on have, on average, slightly higher protein levels than in the forest at large.[9] Wandering in a jack pine forest one day, I saw striking evidence of this ability in the form of a single pine that had been *denuded* of needles and had a mound of spruce grouse droppings beneath. The tree had presumably grown from seed on the spot where a moose had died or some other chance event has occurred, fertilizing the spot to grow a nutrient-rich and tasty pine. Geese can do this too, picking out recently fertilized fields from miles away.[10] It is thought that animals see something in rich foliage, some subtle shading of colour, and then confirm it by taste.[11]

Wildlife have basically the same need for nutrients – carbohydrates, protein, fat, minerals and vitamins – that we do, but can have a devilish time getting them. Perhaps the best example of this comes from our old friend the moose. Each spring, right after snowmelt, moose appear by dozens along roadsides in the Park; it is common for visitors to see 10 to 15 moose on the short drive across Highway 60, the Park's main southern thoroughfare, in mid-May. The sudden emergence of moose coincides exactly with the eruption of flies and mosquitoes (and warblers) and people assume that the moose are coming out to roads to escape flies in the "bush." That assumption is just plain wrong (stand by a road in mid-May in the Park and see how many flies you attract); what the moose are really after is salt, and in particular, sodium.

Moose, you see, live exclusively on woody browse (twigs) from the time snow hits the ground in mid-November till spring. Moose are more dependent on browse than any other member of the deer family, and this close link earned them the name "Twig Eater" in some Native languages. Woody browse is severely lacking in nutrients and contains practically no sodium, a mineral that animals need to balance the water pressure in their cells. Moose, therefore, come out of winter with a severe "sodium thirst" and flock to roadside pools where highway road salt collects from winter.[12] (It is actually something of an enigma that moose can survive solely on salt-deficient browse as long as they do.) The attraction to roadside salt results in the loss of upwards of 15 moose being killed by vehicles every year in the Park.

Moose are huge animals that live most of the year in bitter cold and need to take in extraordinary amounts of energy from food to survive. By means too arcane to relate, researchers have calculated that a large bull moose needs about 17,000 kilocalories of energy a day, more than seven times as much as a grown man, just to meet their basic metabolic needs. To accomplish this moose have to consume huge amounts of browse – upwards of 50 pounds daily.[13]

Were it not for their superb adaptations to the rigors of winter, moose would require a whole lot more energy. Perhaps because we humans are so uncomfortable in cold, it is hard for us to accept that anything could be happy plowing through four feet of snow in -30° C weather for weeks on end. Moose, though, appear to be quite comfortable in winter. The sheer bulk of the animal provides for an energy efficient frame. This, and a thick coat and the long legs to navigate in deep snow, make moose a classic "chionophile" (literally, "snow-lover") – a term coined by snow researcher Bill Pruitt of the University of Manitoba).[14] Moose are, in fact, so comfortable with winter that the onset of a warm spell, the "January thaws" of mid-winter, can cause them significant distress. Temperatures above -5° C in mid-winter are quite discomforting and moose will sprawl on the snow to dissipate body heat much as a dog will lie on a cool basement floor on a hot summer's day.[15]

The really intense heat of mid-summer takes the matter from an inconvenience to potential lethality. The energy-efficient form and massive bulk of moose make it very difficult to dispel excess heat in summer, an absolute necessity to avoid the runaway and often fatal cooking of "heat stroke." Temperatures in excess of 20° C force moose to seek comfort in water or the cool, shaded forest interior.[16] The real "dog days" of summer, the intense heat of 30° C plus that we occasionally see in Algonquin, can be deadly. This intolerance of heat limits the moose to northerly climes.[17] (The brainworm carried by deer also does this because deer are generally more abundant in the south.)

You and I, if lost in the woods, cannot simply turn to munching on twigs; apart from getting the worst imaginable case of the "trots," we would gain nothing for our pains. The reason that moose, and other "twig eaters," can is that they carry within them the means of coping with their incredibly poor diet.

Woody browse is awfully deficient in nutrients but, worse yet, most of what is there is "locked up" within the structure of the twig's cells. To retain their rigid form, woody plants have to have very stiff cell walls. The cell walls are formed by tough, fibrous carbohydrates, principally cellulose and lignin, that cannot be broken up by digestive enzymes. Moose carry in their enormous guts an entire population of bacteria that perform that role. This great mass of bacteria, the so-called "gut flora," enjoy a "happy" relationship with moose; the browser (moose) provides a cozy place to live and the bacteria do the vital service of dissolving the cellulose and lignin trapped within cell walls, thus releasing vital nutrients to the moose. All this has a cost because the "gut flora" of moose and all other members of the deer family take some of the nutrients they release to serve their own needs to feed, grow and reproduce. Without this internal army of bacteria, however, most herbivores could not survive.[18]

Newborn moose calves face a great predicament, for they are born without the vital bacteria in their stomachs and must acquire a shot or "inoculum" of bacteria before being weaned at about three months of age. Failure to do so in time means certain starvation.[19] In the interim, however, calves are on mother's milk (which, incidentally, is way richer than cow's milk) and this milk has to find its way past the growing mass of bacteria or be, literally, curdled. To accomplish this, suckling calves pass their milk through a groove that forms temporarily along the lining of the stomach, bypassing the hungry bacteria.[20]

That stomach, the enormous and convoluted gut of the moose, is a truly marvellous mechanism, one of nature's inspired creations. Moose, like cows, are "ruminants," meaning they must "chew the cud" to extract enough from their meagre diet to survive. To facilitate this, moose and other ruminants have evolved a massive four-chambered stomach, the composite parts of which work in exquisite harmony to pull every trace of benefit from plant food, essentially permitting the moose to survive in its harsh environment. The moose, when not actually eating, is involved almost incessantly with the consequences of eating – regurgitating and rechewing a great mass, or "bolus," of "digesta" in lengthy bouts of "ruminating." Cows can be seen doing this on any summer day's drive in the country and seem to be engaged in a happy but inane ritual. Ruminating is, in fact, absolutely essential, for survival for the fibrous plant matter has to be crushed and swallowed again and again to give the gut bacteria enough points of access to completely dissolve the cell walls. Only when the mass is properly pulverized can it be passed from the first chamber (the rumen) to the other three, which then perform in concert, each with its own specialized role in extracting water, vitamins, minerals and, finally, the amino acids needed to build protein and muscle. The study of such things as the gut of moose is not the most glamorous of scientific pursuits but is vital to understanding the animal and how it interacts with its environment.[21]

Moose are so involved with the exercise and consequences of eating that they do practically nothing else. The challenge of maintaining a massive body on such a pitiably poor diet forces the animal into a monotonous routine of searching for, browsing on, and ruminating food. Studies of the daily "time budget" of moose, in fact, show that the massive animals do little else but eat and ruminate.[22] As such, a moose can be legitimately viewed as a sort of life-support system for a stomach or, conversely, as a mobile stomach driving the whole organism in a relentless search for food.

There is one time of the year that, for bull moose at least, this cycle is broken. Moose breed or "rut" in early fall and the drive to reproduce becomes an all-consuming passion. Bulls become totally committed to the elaborate rituals of attracting and pursuing cows and staving off rivals; they pursue this activity with incredible energy, essentially running amok for about six weeks. A mature bull, thus engaged, will forget food for days on end and when the rut is over will have expended much of its vital energy reserves.[23] These reserves, stored as body fat from bouts of gorging on rich summer foods, are needed to see the moose through the lean times of winter. A bull that has had a particularly amorous fall can thus go into winter at a severe disadvantage.[24] At this point, bull moose, the great and powerful Lords of the Forest are, paradoxically (and excepting small, late-born calves), the most vulnerable of moose and often the first to succumb to starvation in the really hard winters.

White-tailed deer are also ruminants and that fact has frequently caused them serious difficulty due to well-intentioned but over-hasty attempts to help them through winter. Deer, as discussed earlier, are not nearly as well-adapted to deal with the rigours of winter as moose, and frequently suffer serious losses. People throughout the north get together and sponsor "supplementary feeding" programs to help deer through the hard times. These efforts, however, can sometimes do more harm than good due to the exacting biology of rumination. The impulse people feel upon seeing deer in trouble is to get them food – good food, lots of it and fast. Well-intentioned sportsman and naturalist clubs will often put out grain in mid-winter, hoping to give the animals a timely reprieve. The problem is that the rumen, or main gut, of the deer has been functioning at this point for weeks on woody browse, which we know to be very poor feed. A particular populace of microbes is in place in the gut to deal with the natural diet but not the grains that are so rich in starch and other sugars. Deer gorge on the grain but it just sits unattended in the gut and begins to, literally, ferment. Fermentation produces toxic acidic by-products that inflame the lining of the stomach (technically "rumenitis") and can cause fatal blood poisoning.[25]

Cases of acute rumenitis like this are actually infrequent; it is more common to see deer suffering from varying severity of the "runs." One

*Deer in front of the sign at the original site of the Park headquarters – Cache Lake. Today the Park Headquarters is at the east gate.* Algonquin Park Museum Archives #2652 – George Phillips.

nevertheless has to be very cautious in feeding deer in winter. Supplementary feeding of deer has a role in management but must be done carefully; the richer feeds need to be introduced gradually to allow the deer (or more accurately their rumen microbes) to adapt.

We see then, that moose are superbly adapted to maintain their enormous bulk through winter despite their exceptionally poor diet, but just how big, exactly, are moose? Perhaps surprisingly, given all that we do know about moose, that very basic detail of their biology was not exactly clear until quite recently. There is, in fact, a great deal of misunderstanding about the size of moose.

A friend of mine, a travelling salesman, once suffered this misapprehension most personally on a trip to Newfoundland. He engaged in a heated dispute with some local lads in a bar in St. John's over the size of Ontario versus Newfoundland moose. Emboldened, no doubt, by a few tots, he claimed an Ontario weight of about 4,000 pounds to which the Newfoundlanders, sensible fellows, responded with a more realistic figure. My friend, I'm happy to say, although a little worse for wear the next morning, is still with us but his experience underlines a common misconception. Most people considerably overestimate the size of moose.

Moose do appear larger than they really are. Long legs, a deep chest, the overall dark colouration, and perhaps the inspired state of the viewer give the impression of a really massive animal. A truly enormous moose did exist some 200,000 years ago in the mid-Pleistocene. *Alces latifrons*,[26] probable ancestor of our modern moose, is estimated to have weighed up to 1,000 kg (2,200 pounds). Today's version is much smaller but it wasn't until quite recently that we knew the real size and range of weights of moose across North America. This was undoubtedly because of the practical obstacles encountered in trying to capture and weigh such huge animals whole in the field.

Randolph Peterson, Curator of Mammalogy at the Royal Ontario Museum, first examined the subject in 1955 and found that there was virtually no information on whole weights of moose in North America. Peterson reviewed the subject again in 1974 and noted that the picture had not really changed.[27] Through the 1970s and early '80s accounts of moose weights began to appear in the scientific literature just as tranquilizing drugs, and the opportunity they provide to handle live animals, came into really common use. Most of these reports, however, were from western moose, not the eastern (*Alces alces americana*) subspecies found in Algonquin.

An opportunity to weigh a significant number of eastern moose came with the Michigan moose transfers of 1985 and '87, discussed in the last chapter. We weighed all 67 moose captured in the project and followed this up by weighing road kills in the Park for several years. By 1990 we had weighed a total of 89 animals, a significant sample.

Results were surprising. We divided the weights into "young adult" and "mature adult" categories, the cut-off being age six when moose are fully grown (you can make a pretty good guess at the age of moose from an examination of their teeth). The mature adults averaged only 470 kilograms (1,035 pounds) with cows averaging 460 kg (1,015 pounds) and bulls 495 kg (1,090 pounds). The biggest specimen, a beautiful big bull that made the long haul to Michigan, was 540 kg (1,090 pounds). With antlers up he would have gone over 1,250 pounds – impressive but not the stuff of Great Northern Legend or bar room swagger. Moose, it seems, are not quite the mammoths they appear.[28] (The Alaskan version, the moose appropriately tagged *gigas* by science, is, however, considerably bigger than the eastern version and may go over 730 kg (1,600 pounds.)

The environment that moose live in is, in one way or another, severe most of the year and this has no doubt set an upper limit on the size the animal can attain. To survive, a moose must learn to take quick advantage of the different opportunities provided through the seasons in its constantly changing surroundings. In summer, moose move to water and are commonly seen immersed in swamps grazing on water lilies and other aquatic weeds. It is commonly held that moose do so to escape the summer's heat and flies and there is some truth in this popular wisdom. The real reason, however, is the unrelenting drive to seek high quality food and sodium in particular. Aquatic plants are particularly rich in sodium[29] and, as discussed earlier, moose are desperate to sate their need for dietary salt in spring. Water weeds are also particularly high in protein and moose gorge through summer to build up their strength for the leaner times to come.

Moose are surprisingly graceful in water and morph into an essentially aquatic animal in the brief northern summer. The great animals become almost fish-like and have been known to dive up to 18 feet deep to get at the rootstocks of water lilies[30] (imagine the shock of an unsuspecting scuba diver!). Moose can stay submerged for up to a minute and do so by splaying their large hooves, paddling furiously and expelling air. Calves can swim at as early as a week old, but very young calves have been known to drown.[31]

As summer progresses, moose leave the wetlands and seek the cool, dark forest interior where they feed on fresh greenery, often skillfully "stripping" leaves down the entire length of branches. Moose deliberately seek the very densest forest at this time because the plants beneath, being shaded, are still fresh and have not yet developed the toxic by-products of older growth[32] (see below). In late fall, moose gorge on fallen leaves because even dead leaves are more nutritious than the only alternative – woody browse.[33]

Then the snow hits and moose have to subsist for months on a desperately poor diet. The twigs of many types of shrubs may have less than 7 per cent protein,[34] which is just barely (and perhaps not) sufficient to sustain mammals like moose for extended periods.[35] The poorer quality shrubs, like young sugar maple, are (unfortunately for moose) the more common, and moose have to choose between eating large quantities of poor browse or using precious energy to seek out better stuff. The choice can be a life or death decision.

Moose, then, are faced with an unremitting dilemma, constantly having to weigh the benefits of quantity versus quality. This situation is common in wildlife and has given rise to an entire field of study with the imposing name of "Optimal Foraging Strategy." But more on that shortly.

Can plants fight back? Do plants have any defense at all against the relentless assaults of herbivores? Plants actually employ an amazing array of defenses, some of which are ingenious, even fantastic.

When you think of it, feeding on plants is just another form of predation except that the victim is stationary, and plants have had to develop means of defending themselves short of fleeing (something they obviously cannot do). The most common of these is the employment of various chemicals that, if not actually toxic, make eating the plant a losing venture. Many of the shrubs eaten by moose and deer contain high levels of various aromatic chemicals (mostly terpenes and phenols) that make the plant taste bad or actually inhibit digestion. Certain compounds are "bacteriostatic" and thwart the workings of the bacteria in the rumen, thereby preventing the release of vital nutrients from the food.[36] The ingestion of certain winter foods can actually be worse than not eating at all as the energy used to find, consume and digest the browse is more than that obtained from it. Some winter forage is, in fact, so poor that they cannot support even the bacteria in the deer's gut, let alone the deer itself![37] Eating this stuff is obviously a losing strategy and moose and deer avoid certain shrubs, such as alder, that are high in toxic compounds. However, in the pain of hunger anything will do and moose and deer will, if desperate enough, gorge on the toxic shrubs. This can be a fatal choice, especially to the slighter, more vulnerable deer. Deer, incredibly, can die more quickly with full than empty stomachs.

But plants have much more sophisticated tricks. One of these came to light during World War II in Holland where Dutch women, forced by famine to eat tulip bulbs, suffered upsets in their menstrual cycles. Researchers studied the chemistry of tulips after the war and found that the bulbs contained compounds that mimic estrogen, the sex hormone of female mammals.[38] Further study showed that many other plants produce similar agents that are more active than estrogen itself. Thai women, for example, have been reported to use a powerful extract from the *Peueria* tree to induce abortions.[39] It is well established that reproduction in sheep can be impaired when they feed on certain types of clover. Plants can thus disable the reproductive processes of herbivores, driving down entire populations and gaining a fitting revenge for the abuses they endure.

A marvellous example of plants actually *assisting* their oppressors comes from work, by A.S. Leopold, with quail in California. (This A. Starker Leopold being the son of the celebrated forester and naturalist Aldo Leopold, mentioned above in the context of warblers, and author of the classic *A Sand County Almanac*.) Leopold determined that the content of isoflavones (another group of estrogen-like chemicals) in pasture plants fed on by quail varies sharply from wet "rich" years to drought years. In good years, the isoflavones are low, and quail reproduce freely, but in poor years, the isoflavones rise, reducing egg production by the birds and thus increasing the survival chances of the few chicks that are produced.[40] The pasture plants thus help maintain a balance in numbers for the quail although why the munificence exists is uncertain. The favour, no doubt, is unintended.

The ultimate in plant defense strategies (technically "antiherbivory") was first suggested by observations of acacia trees on the African Savannah in the early '90s. Biologists noticed an odd pattern in the timing and movement of giraffes feeding on acacias. The animals would typically feed on one or two trees and then stop and, for no apparent reason, move a considerable distance to feed again, bypassing closer trees in the process. The pattern strongly suggested some sort of communication, a warning, being transmitted by the damaged trees to their immediate neighbours. The idea of plants communicating seemed fantastic, and what was actually happening with the Acacias was never clearly established. Subsequent work, however, established that plants really can "warn" one another. Research in the early '80s demonstrated that poplar and maple saplings that were damaged in the lab communicated the fact by some airborne cue to nearby undamaged plants which responded by increasing the levels of noxious compounds in their foliage[41] (which, in the wild, would presumably deter herbivores). Communication and defense in plants seemed real. Scientists, nevertheless, are skeptical of such improbable stuff and the whole question of plants "talking" to each other has recently been revisited with a more jaundiced eye.[42]

Actually the relationship between herbivores and their prey, plants, has a lot of positive aspects. Herbivores recycle plant material and release the nitrogen needed for plant growth into the soil through their droppings. Herbivores can also influence the pattern of natural fires; grazing by buffalo, for example, can reduce the intensity of fires in prairie grasslands.[43]

As we have seen, however, plants can defend themselves.

Moose, then, are compelled by vital necessity to adjust to nature's irregularities but perhaps the most intriguing demonstration of "optimal foraging" comes from work in the Park with another symbol of the north country, the beaver.

John Fryxell is another among the long list of outstanding biologists to come to the Park from the University of Guelph. A tall, lean man with the thick dark beard that is *de rigueur* among field biologists, Fryxell chose the beaver to test whether animals use any sort of "conscious" strategy while foraging or just eat whatever they come across. A relative newcomer to the Park, Fryxell seems to be settling in, perhaps to add his face to the cookhouse gallery alongside Anderson, Pimlott and the like for years to come.

Beaver are, like moose, browsers most of the time, the big difference being that they have to cut trees down first to get at them (and yes, beaver do occasionally fell trees on themselves – with fatal consequences). Beaver, also like moose, feed heavily on aquatic vegetation, and particularly water lilies, in summer. Beaver and water lilies are a perfect example of collaboration or "commensalism" in nature. Water lilies feed beaver, but the ponds beaver build provide ideal habitat for the lilies to grow in. Beaver also sow water lilies about the ponds by scattering rootstocks while feeding, in essence farming the lilies.

By fall, however, beaver have to turn to poorer food and, like moose, become twig eaters. Fryxell wanted to see whether beaver show any sort of "strategy" in feeding to optimize their intake of food (and odds of surviving to produce young) as theoretically they should. The models he used have foot-long equations to predict how predators (in this case the beaver) will use time and energy in pursuit of prey (the twigs) more efficiently as they range farther and farther from home. Fryxell and his students conducted a set of experiments by confining beaver to large pens with artificial ponds, planting twigs in strategic patterns, and then painstakingly recording how the beaver went about feeding.

They had to work at night because, like many herbivores, beaver feed mainly at night. There is actually a great deal of night work at the station, for wildlife are, in general, more active by night than day and researchers have to adapt to the lifestyle of their subjects. Dawn is also a very active time, particularly for songbirds, a major study group, and the sun rises extraordinarily early in the North in late spring when the station is busiest. All these comings and goings, and the natural tendency for young people to indulge in serious frivolity, adds confusion to an already confused scene. It is common on my frequent stops to cadge a lunch to see one group of academics coming off or going on shift, staggering in unshaven and glaring at their rival scholars who have kept or will keep them up all night or day.

The most outrageous manifestation of this group rivalry at the station occurred many years ago and the story is true although it is one of those for which the principals have to remain nameless. It seems that a group of parasitologists (who, as we have seen, are an eccentric if not outright perverse lot) were studying roundworm parasites of bears. Parasites, being mostly internal, have no great sense of day or night, so this group was natural for the day shift. The bear people had been tormented for weeks by a raucous group of night owls (who were, in fact, according to the story, studying real owls). The bear research required anaesthetizing the animals and, lounging about one sunny day after working on a large adult, one wag hit on the idea (you've probably guessed where this is going) of putting the unconscious bear to bed with one of the ornithologists. This they did, but not before dressing the bear (a female) in a bonnet and bra. This seemed a ripping good idea but the hapless owl expert did not, as planned, wake up. After some time the bear people began to fret that the drug would wear off and the whole affair end without humour. The prank ended in predictable chaos as the parasitologists struggled frantically to extricate the now semi-conscious animal from the tight quarters of a hysterical Associate Professor of Ornithology, who had finally awoken and by all accounts has not been the same since. All this, incidentally, occurred in June, when bears mate, and the animal was probably in heat.

As a sign of how things have changed, a prank like this would be unthinkable nowadays, the sensitivity around handling of animals in research being extreme.

The modern-day night shift, Fryxell's people, perched on makeshift platforms and stared like bug-eyed watchful owls into the dimly lit night for hours, meticulously recording every detail of the beavers' activity. The students returned by day and measured the effects of the unwitting beavers efforts on the ground, the pattern and intensity of browsing. Results of all these idiosyncratic labours were at the same time predictable and unexpected. For example, the number of saplings the beaver ate decreased with distance from the lodge; beaver were, not surprisingly, eating close to home first. Less obviously, Fryxell found that beaver ate larger and larger twigs the further from the lodge they roamed.[44] This seemingly trivial observation has (with thought), some serious implications for how we may perceive of the animal's "thinking." The beaver, recognizing that they are farther from home, apparently begin "consciously" to choose larger food items. This makes sense, because as the animal invests more energy searching for food there has to be a payoff, a better return on investment, or the exercise becomes self-defeating. But how do the animals actually perceive and adjust to their circumstances?

We humans, in a similar situation, would react mindfully – "Right!

I'm getting farther from base, better start picking the bigger bananas or the whole trip is a frost" – but does anyone really imagine that animals think that way? (If they do we have to confer on them a higher station than most of us, I expect, would accept.) Such behaviour must be innate, that is, in the genes, but it is hard to accept that behaviour that seems to involve "choice" is nothing more than chemically-driven mechanics. The whole question of animal behaviour and its causes and complexities is a fascinating one, one that drives Fryxell and his ilk to do things like sit for hours in the cold, the dark and the bugs to unravel the gustatory manoeuverings of beavers.

The tactics beaver employ in feeding may be more than just a matter of efficiency. Beaver are extremely vulnerable to wolves on land and place themselves at great risk when far from water. Optimizing time spent near home is probably as much a matter of survival as it is good management.

In the course of his work and for reasons too obscure to explain, Fryxell had to figure out the food retention time of beaver or, more to the point, how often beaver sh – . The standard means of doing so is to have an animal consume some sort of marker with food, in this case, coloured beads. Try as they might, though, Fryxell and his students could not coerce the beaver to eat the beads. Could not, that is, until one resourceful fellow devised a sort of "maple sandwich" – a doughy paste containing the beads sandwiched between two "slices" of maple leaf. It was through this device, which the beaver relished, that Fryxell made the strange discovery that beaver defecate only at night! Why this is the case will likely be known forever only to beavers. My guess is that it is a means of somehow avoiding predators. The discovery, at any rate, necessitated adding someone to the night shift, the poor fellow having to wake up every two hours all night long to check if the beaver had cra – – . He became the "butt" of a great many jokes.

As noted, biologists certainly do get up to some strange things and have a well-deserved reputation for eccentricity, several examples of which we've seen (or will see) in this book. My most memorable experience with this occupational hazard occurred at a conference I attended in the early 1980s.

The meeting was an annual affair – a gathering of North American wildlife specialists that is held in a different location every year. That particular year it was Ontario's honour to be host, and I and a great many of my young colleagues were thrilled to attend. There is, as is the way of such events, a keynote address always delivered by some luminary figure in wildlife research. The speaker that year (who will remain unnamed) was a professor of zoology at a prominent university and a world authority on a certain well-known large mammal; his being such a God-like figure, we were all very eager to hear from him. Our anticipation turned to disbelief when the fellow strode purposefully into the conference hall of this major Toronto hotel dressed, literally, in rags. I say literally without exaggeration for he had on a strange sort of tunic that consisted (I'm not making this up) of household rags stapled together. The man sported a dishevelled ponytail (this being before they were fashionable in men) behind a scarred, craggy face strikingly suggestive of Willie Nelson on a bad day. He proceeded to deliver a superb talk, at the completion of which he received an enthusiastic ovation.

I had really wanted to meet the man and, inspired all the more by his eccentric appearance, went to seek him out at coffee break. I found him chatting amiably with a group of admirers all of whom were spaced an exact six feet back, forming a perfect circle as with a wagon train around a campfire. The reason for this became apparent when, blundering ahead, I was stopped short by an appalling stench that hung over the fellow like an aura. The man smelled like a goat and not one of us could get near him. I assumed that, as some means of bonding with nature, he never washed. This is, you understand, from a man whose income, even then, had to be in the six-figure range.

I know of another wildlife research biologist who was so dedicated to his work that he rarely went home, essentially living in his office. When the fellow died they found dozens of paycheques in his desk; he had not even bothered to cash his cheques. The profession, like other disciplines of science, seems to attract such obsessive personalities.

Beaver, at any rate, are getting scarcer in the Park. Like deer and moose, beaver thrive in young forests, which provide plenty of shrubby regeneration for food, particularly of their favourites – the sun-loving aspens and birch. Beaver flourished in the Park in the 1930s and '40s (the heyday also of deer) because extensive logging at the time left thousands of acres of this "second growth." There are accounts from the late '30s of hundreds of beaver dams and houses within just a few miles radius of the general vicinity of the Wildlife Station.[45] Since then, beaver have been on a long, slow slide and there are now probably fewer than half the number of beaver in the Park than at their peak. This, in truth, is entirely our doing.

The problem is that we have simply become too good at putting out forest fires. Fires, apart from their obvious destructive potential, rejuvenate the forest, creating "pastures" of aspen, and when permitted to burn near swamps and creeks make excellent habitat for beaver. We have been very effectively suppressing forest fires in the Park now for decades and the result is that the forest is maturing everywhere, to the detriment of beaver, deer and, to a lesser extent, even moose. Logging can compensate for this to a degree, and logging is permitted (and carefully controlled) in about two-thirds of the Park. Here again, though, good intentions may, at least in part, be going wrong. Logging along shorelands, and particularly trout waters, is restricted with "reserves" to prevent erosion. This, of course, is a good and necessary thing, but the long-term suppression of "disturbance" along creeks and rivers is making things tough for beavers which, as we have seen, need their food close to home.

The apparent decline of beaver is of great import to the entire forest community for beaver are, like moose, a "keystone" species. Keystone species are those whose activities, in this case the creation of dams and ponds, significantly alter the environment, creating (or destroying) habitat for many other species. (Another good example of a "keystone" effect is the famous plague of introduced rabbits in Australia which completely transformed parts of the Outback.) There are options to improve the lot of beaver in the Park. For example, shorelands could, with no effect on trout waters, be partially logged where the surrounding terrain is flat and forest fires could be allowed to burn in certain weather conditions (concerns for public safety being, of course, paramount). These options are being examined by forest managers in and outside the Park in ongoing efforts to balance logging of the forests with what nature intended.

All this, incidentally, is not to suggest that beaver are in any sense rare in the Park. Beaver are still common and, being as resourceful as their reputation, will always find places to scratch out a living. It may be time, however, to give them a hand.

Perhaps the ultimate in ingenuity in finding food is demonstrated by the black bear. There is, contrary to popular belief, only one species of bear in the forests of eastern North America (the confusion apparently resulting from the occasional appearance of brownish variations – these are just brown "black" bears, not the much larger true brown bears of the West, of which grizzly bears are a variety). Bears are famous hibernators and have to put on great reserves of body fat to sustain them through long months in the den. To do so, bears go into an orgy of eating in late summer and fall, seeking in particular foods that are high in energy and fat. These are mainly various forms of "mast" – the fruits and nuts of trees. The most important of these, at least in the type of hardwood forest we have in Algonquin Park, is beech nuts.

The American beech is the familiar massive hardwood tree of farm woodlots with the striking blue-grey bark. Beech trees bear nuts every fall but every few years produce a bumper crop, and bears gorge on the fatty nuts through the brief northern Indian Summer. Beech nuts are highly nutritious, but getting at them presents a formidable challenge; the nuts are born high on the lofty tips of the highest branches of the great trees. One of the truly outlandish spectacles of the north woods is the sight of a 140 kilogram (300 pound) black bear perched 80 feet high on an impossibly thin branch reaching gingerly for tiny nuts just beyond arm's length. Bears break and fold branches inward to get at the nuts leaving tangled masses high in the trees, the "bears' nests" familiar to woodsmen (which, of course, are not "nests" at all). Black bears are amazingly agile in trees but must occasionally fall to their deaths during these arboreal antics – although such an event has never been recorded.

Bears concentrate their efforts on individual trees, a fact attested to by the common observation of the soft bark of certain beech trees being severely scarred by claw marks from repeated visits while its neighbours remain unscathed. (This focus on individual trees is reminiscent of the denuding of individual pines by grouse referred to earlier.) Studies have shown that the favoured trees, not surprisingly, have the richest supply of nuts. Bears seem to know without climbing which are the best trees and don't generally waste time exploring poor beeches, but how they acquire this knowledge *a priori* remains a great mystery.

Ernest Thompson Seton used beech crops as a portent of winter severity and bear denning activity noting that "if … there has been a good year of mast, and the winter is a mild one (and it is a fact that, with us, good beech nut years are commonly followed by open winters), the male bears prowl about nearly, or quite, all winter."[46]

One should not assume that any bear "sign" scratched on trees is the result of feeding activity. Bears have a year-round affinity for trees, routinely clawing and biting into bark, presumably as a means of marking territory. Seton remarked on this as early as 1909 and on how bears were such great arborealists in general, pointing out that many a hunter, upon missing a shot at a treed bear, is astonished at the speed with which the animals make ground, often with a single bound from great height.[47]

Black bears, at any rate, have a nose for food that is perhaps unmatched by any of our northern fauna. Seton remarked on how in Manitoba they were known to feed on the foul piles of dead and rotting mayflies that wash up in great windrows along lakeshores in early summer. As such, their only equal may be the Raven which seem to appear almost as from spontaneous generation wherever food appears. They actually do this by scouring the countryside from the air using roosts as "communication centres" to pass the location of food along rapidly. The extraordinary capacity of ravens (and crows) to find food, especially carrion, is a skill exploited grimly by experienced woodsmen when searching for missing people; find the ravens and you, often as not, find the body.

Fryxell is not the only biologist in the Park who has been interested in the strategies animals use to optimize their food intake. An elegant little study of the same question, but on ovenbirds, was conducted at the station in 1974 by Reto Zach and Bruce Falls of the University of Toronto. (Professor Falls, now retired, is another of the Wildlife Station's venerable pioneers who, among other things, started what has become the longest continuous study of small mammals in the world.)[48]

Ovenbirds are inconspicuous, brownish, ground-feeding warblers. The tiny birds are plain in appearance, but have two quirks of behaviour from which they have gained some distinction. First, they nest on the ground in little "ovens" that they fashion artfully from bits of grass and debris and, second, they have perhaps the most conspicuous call of our spring birds, a strident TEACHER – TEACHER – TEACHER that rings loudly for weeks in spring and summer through the deep forest interior. Ovenbirds are easy to tame and handle and this, and their ground-foraging habits, made them ideal subjects for study.

Zach and Falls confined ovenbirds to a 6 metre by 6 metre seminatural feeding "arena" on the forest floor on which they placed grubs in different densities in 1.5-metre squares of a grid. They wanted to see if the birds optimized their time and efforts by foraging where the grubs were thickest or just wandered about randomly. They let the birds loose and, like their counterparts studying beaver, retreated to a nearby blind to record their subjects' activity for hours on end. The ovenbirds, perhaps not surprisingly, concentrated their feeding in zones where the grubs were thickest and adjusted the "tortuousity" of their search paths to optimize their intake of food.[49] More importantly, the birds showed clear evidence of having memory, an ability that even at its crudest level, has only been shown for a few animals. Zach and Falls drew parallels with studies of Carrion Crows of Europe (the crow family or *Corvidae* being considered the most highly evolved and intelligent of birds). The crows had shown clear evidence of a well-developed and lasting memory for places at which they have been rewarded. Such abilities, though seemingly basic and obvious, are not trivial. Moving around, you see, is a risky business; locomotion consumes a lot of energy and exposes the animal to increased risk of predation and must therefore be done as efficiently as possible. Zach and Falls then made another clever link, this time between their ovenbirds and a type of highly "tribal" monkey in Africa. Red Colobus monkeys, it seems, move about the jungle with a "cropping rhythm" timed exquisitely to visit food sources at intervals that exactly match the time each plant needs to recover from previous assaults.[50] Nature's strategies, it seems, are universal.

What, you might legitimately ask, is the "value" of such research? There is, in truth, little practical return from these sorts of studies, but people like Fryxell, Zach, and Falls are driven to unravel the workings of nature and many more will follow in their path. The work with beaver and ovenbirds is part of a rich field of inquiry known as "behavioural ecology," in which researchers probe nature's schemes and devices, seeking universal blueprints for survival for the pure fun of it. We are a fundamentally curious species and the volume of completely "unapplied" work done out of Algonquin's unassuming wildlife station is a testament to that fact.

In truth, though, almost any work with important species like beaver might someday have practical application. Beaver were historically, and remain to this day, the mainstay of the fur industry and their workings pose a serious threat to highways and property almost everywhere they are found. Any knowledge of the animals' biology has potential "management" application. There are people at the Wildlife Station studying how the larvae of blackflies spin webs (there are, incidentally, more than 30 species of blackflies, one of which feeds exclusively on loons, as will be seen in a later chapter). Who knows what they might unearth about the twists and turns (literally) of blackfly biology that might someday lead to measures to control this great scourge of the north?

Food, along with water, cover and den sites, is really just another component of habitat, and habitat is perhaps the most devilishly difficult thing for biologists to understand. This is because most wildlife habitats are exceedingly complex and our attempts to measure habitat often miss the point, it being impossible to project ourselves into the minds of animals and see what it is they see. The "niches" animals seek are just too elusive for human perception. An excellent example of this is the way that birds, warblers being a good example, return to stake a territory in *exactly* the same spot in the complex, confused structure of a forest year after year.[51] This happens even after individual birds die; newcomers take their position precisely, even to the use of the favoured singing perch. Researchers viewing the forest have great difficulty figuring out just what it is that makes these spots "just right." One has to be a warbler to appreciate the nuances of avian interior design.

# 9 | Moose Days and Jays

*I never saw a wild thing sorry for itself.*[1]
— D.H. Lawrence, "Self Pity"

In the spring of 1977, I was attacked by a moose. The fact of that, actually, is quite a distinction because moose, being great docile beasts and secure in their domain, rarely attack people. Moose simply have little cause to be aggressive.

There are, however, two times of the year that moose are dangerous — in early fall, when rutting bulls will attack, and in late May, when cows will aggressively defend their newborn calves. Even at these times, though, moose do not normally present a great hazard. There are, to be sure, some risks; it is said that getting between a cow and her calf in spring is almost suicidal. But that is a considerable exaggeration and as proof consider the fact that biologists have studied cow and calf moose in the Algonquin Park for more than ten years, frequently interceding between the two and, although there were dozens of bluffs and false charges, and a good many stories to tell of both, no one was ever hurt. Fatal attacks by moose on people, although not unknown, are extremely rare.

Just my luck, then, to be the object of one. We were studying spruce grouse in New Brunswick, our objective being to catch and band as many of the birds as possible to determine the size of the population.

The work involved long hours of tracking down the grouse with trained pointing dogs. I was walking a narrow trail in a pine woods one day when my dog, a puckish English Setter, which (as was her annoying habit) had gotten far ahead of me and out of sight, started barking furiously. Within seconds, the dog was tearing back down the trail in my direction pursued by a great big and very angry cow moose. The dog, to my everlasting annoyance, ran right past me and out of sight down the trail, leaving me to face the moose which was approaching with disconcerting haste. We were in a jack pine stand and jack pine, you may know, are a spindly, insubstantial tree – not much for taking cover in. I nevertheless scrambled up the stoutest one I could find, perching as high as I possibly could on a couple of slender branches that gave indication of snapping at any moment.

There followed what I suppose to an observer, had there been one, would have been a prolonged comedy. As I swayed precariously at the tip of the pathetic bush, the cow was positioned only a few feet below, giving every indication of wanting to tear me apart, her eyes glazed and red with anger and the hair on her shoulders sticking straight up. The moose would occasionally back off and seem about to leave when, with a nod of her head that seemed to restore her resolve, she would return to assault my tree. This went on for hours. Getting hungry, I was seriously considering climbing down and taking my chances when the cow finally left. The cowardly dog, which had been hovering surreptitiously at the edge of the scene, appeared almost immediately, tail wagging, and apparently feeling entirely proud of her role in the affair.

I left the scene contemplating the vagaries of animal behaviour. I knew something about moose and was certain that the cow had a calf nearby (it was early June) and was simply defending it. I was also certain that it was the dog that had triggered the attack and the moose, seeing the dog escaping, had transferred her anger to me.

I had, you see, just finished a course on animal behaviour at the University of New Brunswick and had learned something also about "releasers."

The study of animal behaviour is perhaps the most "unscientific" of the biological sciences, because it requires the observer to place his or her self in the "mind" of the animal, and it is really hard not to attribute human motives to the subjects being studied. It is also one of the most challenging because it requires intuition and creativity that are uncommon in scientists. This is particularly true if one is studying animals in the wild.

Animal behaviourists ("ethologists") are generally of the opinion that animals live at a level of consciousness or awareness that is far below our own. According to this view even "higher" animals are simple automatons, barely aware of their own existence. Animals are just reacting to events, just as you and I yank our hand off a hot stove element without thinking, and nothing else. There is no room for anything like reason or decision making. Even apparently complex behaviour, like a dog caring for its pups, is nothing but a string of continuous, automatic, reactions to stimuli or "releasers" – in this case the pups, their appearance, scent and whining.

A releaser is a stimulus or signal that is received by the senses, generally vision, and evokes or "releases" a stereotyped response. The sight of a female peacock, for example, will release the glorious tail display from a male. My dog was a releaser to the moose but the behaviour that was "released," an attack, was misplaced; it was really intended to apply to a wolf.

As docile as moose generally are, certain things can release aggression. Being small, or at least short, is a fundamental risk factor. Moose just don't like things that, like wolves, are low to the ground. This came to light during work Ed Addison was doing with moose in the Park in the early '80s. Ed and his students were studying young moose confined to pens at the Wildlife Station. The work involved detailed observations of behaviour and, working with moose in and around the pens for weeks on end, the researchers developed a certain sense of what made them tick. One impression that developed almost subconsciously was a sense that being low to the ground, in a crouch for example (something the team learned not to do), was the only thing that seemed to make them mad.

This impression was confirmed dramatically one day with the only event that the research team (who worked closely with moose for years) would ever recall as a real attack. A group of high school students had been recruited to help with the project, doing the routines of feeding and maintaining the pens, etc. Ed had observed that young girls worked much better with the animals than boys, their innate maternal skills seeming to have a calming effect on the moose. Girls were therefore almost always given the task of working in close with the moose. Things were going normally this day until a particularly small 17-year-old girl stepped into the pen. To the astonishment of everyone present, one of the moose instantly rushed her, passing by her taller companions, and pinning the terrified girl in a corner. The moose was a yearling but nevertheless capable of doing damage, and the team had some anxious moments before they were able to extricate the shaken but unharmed girl from the pen.

Researchers have also tried to use releasers to explain bear attacks. The idea is to identify some feature or object that is common to the scene of bear attacks, an element or releaser that triggers aggression, and then eliminate it. The three young boys that were killed on the Park's east side in 1978 were fishing, and actually had a stringer of trout. This, as discussed earlier, set off a flurry of speculation that the fish, or food in general, was the trigger but a review of the record of other attacks found no such common thread.[2] As discussed earlier, we now know that black bear attacks are, plain and simply, predation; the bears are hungry.

Moose, in reacting so robotically to anything low to the ground, are tending to support those that hold the less inspiring view of how animals "think." Moose (some moose at least) can't seem to discriminate between objects as fundamentally different as a wolf and a young girl. Both have the low profile and some other shared features that are sufficiently threatening to release a charge. It is hard to accept, particularly to those of us who keep animals, that the wonderful, complex behaviours we observe are so mechanistic, so soulless. (In truth, there is an emerging school of thought that some animals, in particular higher mammals like chimps, are capable of a sort of reasoning.)[3] For the most part, though, animals seem to live in a rather confined consciousness.

But one should not assume from all this that the study of animal behaviour is a dull and uninspired calling. On the contrary, some fascinating work has been done in this field and none better than that of a colourful, brilliant and exasperating curmudgeon who worked with moose in Algonquin Park.

Anthony B. "Tony" Bubenik was obsessed with moose. Bubenik dissected every detail of the biology and behaviour of moose in a lifetime of study that began in his native Czechoslovakia in the '30s and spanned nearly the entire century. (He died in 1995.) Very few people have come to know and understand another species on this planet as well as Tony did moose. He was one of the most the most honoured wildlife biologists of his time.

Tony Bubenik was almost lost to science early in life. Serving with the Czech resistance in World War II, he had several close brushes with the Nazis and was not a favourite of the Czech regime that followed. Tony's early demise would have been a great shame, for the man is responsible for much of what we know about the behaviour of moose and indeed the deer family in general. Some of his best days were spent in Algonquin Park.

Short and heavy set, Tony sported an austere crewcut and thick wire-rimmed glasses, and had a generally gruff tone about him. This, a heavy Czech accent, and his formidable reputation could be somewhat intimidating. Tony crossed the Atlantic in 1970 but never lost the formal customs and affectations of a Moravian country gentleman and scholar. He went into the field dressed in a Tweed jacket with ornately carved bone buttons. At the end of every session with moose (some of which were, as we shall see, quite exciting) he would produce a flask of Grand Marnier from his vest and insist that everyone toast the event – some sessions concluding just after daybreak. Tony had come to North America to teach biologists to think of moose, and wildlife in general, as something more than just things to count and "manage," but rather as wonderfully complex and sensitive organisms intimately connected to their environment. He was frustrated to the end of his career with our informal, laid-back, ways and would occasionally let it be known. He was, nonetheless, revered by the small and close-knit community of moose biologists from Canada and the northern U.S.[4]

Bubenik's great obsession was with moose behaviour and how the magnificent animals maintain social harmony. This aside, there was absolutely nothing about moose that escaped his attention. For example, he collected dozens of hooves of moose from hunters, studied every detail of their structure, and wrote a great treatise on moose hooves, describing in meticulous detail every nook and cranny of their morphology.[5]

Bubenik was a great observer of detail. One day, Mike Buss, a colleague who worked with Tony for years, was driving him along a side road in the Park when they came across an friend of Tony's who was visiting from Poland. The fellow was staring fixedly at a set of moose tracks at the side of the road. Tony got out and explained (he spoke six languages) that they were the tracks of a cow. Appraised of this in English, Buss was mightily impressed for, although having considerable experience with moose himself, he knew of no way of telling the sex of a moose from its hoof print. Buss went from impressed to astounded when Bubenik proceeded to announce that it was in fact a pregnant cow, and roughly how far along the pregnancy was! (The trick, apparently, had to do with a subtle splaying of the hind hooves under the additional weight of the fetus.)

Much of Tony's work was done for the sheer joy of it and never acquired any practical application. Oddly enough, the business with moose hooves did. The hooves of moose, it turns out, are constantly worn away by the rough terrain and have to be replaced with new growth. Normally, the repair compensates for the wear and the outer structure of the hooves is maintained. However, when moose get really busy, as in the rut in fall, the furious activity wears away the edge of the hoof faster than it can be replaced. The hooves, particularly those of bulls, which are much more active than cows in the rut, then become noticeably damaged. In situations where a moose population is "socially imbalanced" (which was Bubenik's great theme) – for example, where the numbers of moose are very low and the availability of cows is poor – bull moose can have a terrible time. They are forced, in desperation, to search frantically for cows for weeks on end. As we have seen earlier in an earlier chapter, this can consume huge amounts of energy and put the animals at a great disadvantage going into winter.[6] Tony was deeply concerned about the well-being of his charges and hated to see situations like this develop. He developed a sort of index of hoof wear, a chart showing hooves in varying states of deterioration,[7] that wildlife managers could use to measure the social health of their moose herd and act to fix things (reduce hunter quotas for example) if necessary.

Bubenik's powers of observation were legendary. He once engaged the assistance of a moose biologist from Northern Ontario, Tim Timmerman, in collecting moose heads to X-Ray for a study of the bones of the skull (the obscure reason for doing so will remain known forever only to Tony). Timmerman obtained the skulls and, having a connection at the local hospital, had the radiographs done and proudly presented the results to Tony. To Timmerman's great discomfort, Bubenik stared long and hard at the screen without comment, having sensed immediately that something was wrong. There was some imperfection in the images, but he couldn't quite put his finger on the problem. Timmerman protested that all was well, but Bubenik persisted until Tim sheepishly admitted that his dog had got at the moose heads on his garage floor and chewed off the noses! The incredible thing about this was that the dog had eaten only the flesh and, as far as Timmerman could tell, had not touched the bone (which, of course, was the only thing visible on the X-rays). But Bubenik somehow detected a defect and berated Timmerman kindly with "Oh, such a biologist!"[8]

Timmerman is an accomplished moose biologist in his own right and collaborated with Bubenik on another arcane enterprise – a study of the moose "bell." Moose carry an odd, pendulous piece of hairy flesh beneath their throat, popularly known as the bell. The appendage varies in size and shape but is normally about a foot long and larger in bulls than cows but, oddly enough, tends to shrink with age. (Reader, I am not, believe me, venturing into eroticism here.) The shrinking with age is thought to be simply the result of repeated frostbite. Biologists have speculated about the function of the bell for many years, a popular theory being that it served as a "decoy" for attacking wolves to snap at. The patent silliness of this idea aside, it has been rebuffed because wolves in attack mode generally approach from behind, working away at the hindquarters. Timmerman, under the tutelage of Bubenik and

Murray Lankester of Ontario's Lakehead University (who worked on brainworm with Roy Anderson), dissected dozens of moose bells and, using the finest techniques of microscopy, looked at the types and arrangements of cells throughout the appendage. They had suspected the bell was some sort of gland that emitted a chemical signal involved in mating, but Timmerman's work turned up nothing unique about the structure of the bells, they appear to be composed of normal tissues. The bell is simply hanging skin.[9] (As it turns out the bell is involved in mating; bull moose rub the bell in their urine, saturating it with alluring scent to attract cows – but more on that later.)

Bubenik was actually an endocrinologist (one who studies the body's glands and hormones) by training and he tried to view the animal as a whole – from the inside out. He was constantly preaching about a strange and difficult concept known as the Umwelt. (The word comes from the German "Um" – for around – and "Welt" – for world – i.e., "around-world" or "surroundings.") Bubenik introduced hundreds of biologists to the idea, droning on ponderously for hours about the "OHMMVVVELT" in his thick Czech burr, but I doubt that one in a hundred really understood him. The Umwelt always struck me as less a matter of science than of New Age fantasy, but with time I have come to appreciate the idea.

Let me, then, try to explain. The Umwelt is the state of connectivity between the moose and its surroundings. The moose is in its Umwelt when it is in its proper niche and feels in harmony with its environment. The Umwelt is thus a concept that views the animal and its surroundings as a single entity when those surroundings are right for the animal.[10] There is a popular expression that I think nicely describes the state of a human in his/her Umwelt; such a person is said to be "in a groove."

I am, to this day, not sure I fully understand the Umwelt, but I know I'm in mine sprawled in front of my TV on a Sunday afternoon with a beer in hand watching the Buffalo Bills play football (I'm in it, that is, if the Bills are winning). You get the idea.

As fuzzy as the Umwelt concept is, it was pivotal to Bubenik's thinking. Bubenik knew that moose were constantly seeking to position themselves in the Umwelt but that the Umwelt itself changed with the seasons. He also knew that the world moose live in is never as complex as in the fall of the year when the animals enter an interlude of frenzied activity, a period so fervid, so intense, as to verge on madness.

This is the rut, the breeding season of moose, when the north woods echo with the roar of bulls, the moans of cows and the crack of antlers in combat. Moose are in their glory in the rut, indulging for weeks in a vainglorious orgy of narcissistic display, mating and fighting.

Bubenik knew that these frivolities had a serious purpose for they set the social order and well-being of the entire population. He made the study of the rut, and all its strange goings on, the centrepiece of his life's work. He did the best of this work in a secluded corner of Algonquin Park, an open bog near Cameron Lake called "Peter's Pond" – after Peter Smith, of whom we shall shortly learn more.

Bubenik built an observation platform at the bog, a place where bulls dug their "rutting pits" and fought great battles, and together with a sundry crew of collaborators and hangers-on, spent several fall seasons probing and manipulating the strange world of the rut.

We humans, once reaching puberty, are more or less continuously sexually active. Moose, deer and others of their kin have a very different experience of sex, a sort of "recurring puberty" called the rut. As the days get shorter in fall, the level of sex hormones in moose rise and the animals are transformed from the docility and torpor of late-summer to discharge the bizarre rituals of moose sex.

Bull moose rampage through the bush, snorting and roaring madly to attract mates and rivals. The roar of a rutting bull is an awesome, sonorous bellow very similar (I have heard both) to that of a lion. Shattering the calm of a misty September dawn it can, to naive ears, be truly terrifying. Bulls change to a more subtle mode when approaching cows, emitting weird "hiccups" by inhaling sharply. Cows, no shrinking violets, respond by moaning and wailing sporadically, the moans decreasing in frequency but intensifying as the actual act approaches.

Much of this activity is driven by scent, as moose enter a world we can scarcely imagine, a world ruled by the perfumes of sex. Moose have a sort of second nose, a large cavity at the roof of the mouth known as Jacobson's organ that greatly increases their ability to detect scent (but, of course, are no pikers in the nose department, being well known for their pendulous snout). Bubenik, in another of his off-beat inquiries, figured out that the "conchs" (bones in the nose that support scent cells) connected to this organ have a surface area of 830 square centimetres, putting them well ahead of dogs, even bloodhounds (we, for your information, have 4 to 5 square centimetres of conchs).[11] By virtue of this capacity, moose in fall become, in Bubenik's words, "awash in a sea of musk."

The main source of this pungent musk, this "eau de moose," is urine, and rutting moose go to great lengths to saturate themselves in the stuff, essentially dripping with pee from head to foot. Bulls build great pitholes in which they urinate profusely, creating a semi-liquid goo of aromatic mud that they then wallow in with vigour. The real trick, though, is to get urine on the antlers, because antlers thus "perfumed" carry the scent high, to be wafted away on the breeze to receptive cows. This, when you think of it, is a challenge because antlers and the source of urine are at opposite ends of the moose. Moose lower the antlers to facilitate this but the penis of a moose (unlike that of some deer) cannot be pointed in quite the right direction to reach them. Moose get around this by lowering the antlers into the pit one at a time and stomping expertly with their front hooves to squirt urine-soaked mud on the flat surface of the antler palms. Thus prepared, the bull, salivating profusely (saliva is another source of musk) and hiccuping all the while, goes a-courting.[12]

Cow moose are not immune to these wiles and, if in the "proceptive phase," (the point at which they will accept a bull) respond enthusiastically. The cow, as she approaches the actual act of mating, becomes "crazy" (Bubenik's term) to keep in contact with scent from the bull's urine and saliva. From initially just tolerating the bull's advances she progresses to smearing herself with urine and saliva from the bull's bell and neck to urinating and wallowing in the pit itself. The cow attempts almost frantically to envelope and maintain herself in the bull's "aura." As things develop further she will even drink the bull's urine and, should he falter in foreplay, stimulate him to pee by head-butting or "bouncing" his penis.

All this reaches a weird culmination in open country, our northern tundra for example, where scent is a "releaser" that can carry for miles. A bull moose on the tundra may thereby attract an entire harem and the frustrated cows must then compete for his attention. They do so by prancing around the bull, recumbent in his rutting pit, for hours in a bizarre ritualized merry-go-round that Bubenik, showing a poetic side, called the "Dance of the Odalisques" (odalisques being the harlot dancers in a harem).[13]

All of this amorous activity is not, of course, without a purpose. The ritualistic exchanges and mutual immersion in the aromas of lust serve to reinforce the progression of bull and cow to actual mating and synchronize the ejaculation of the bull precisely with ovulation by the cow. It is said (in all seriousness) that the cow gives a special moan at the culmination of the event.

It is this immersion in the world of scent that Bubenik believed accounts for the infrequency of moose attacks. Prime age bulls have practically no natural enemies and, in Bubenik's view, "have absolute physical and probably mental superiority among other species of their social world. When attacked, a fully mature and experienced bull or cow moose does not fear any other species …."[14] This may explain why moose are so reluctant to yield to moving objects like wolves, automobiles and trains. That said, there are definitely threats – bears, as we have seen, present a real danger to calves, and moose will attack if given good reason. Bubenik believed, however, that there had to be *both* a releaser of scent and sight to trigger a moose attack. He based this on his considerable experience working close in with moose in which it seemed they were ready to charge but were trying to get in position to confirm, by scent, the need. In Bubenik's view, the profile of a man is sufficiently similar to that of a bear to trigger an attack but without the smell of a bear, it rarely happens.[15]

The whole business of the rut and its strange goings on is, of course, about selecting a mate. The use of scent is part of a repertoire of behaviours and signals that bull and cow moose use to attract the "fittest" mates. The aura of scent is, as we have seen, a potent stimulator and cow moose almost certainly select bulls at least in part on the merits of the piquancy of their urine. There is, however, another powerful releaser of sexual choice and that, of course, is antlers. Moose antlers, Tony Bubenik and Algonquin Park came together in the late 1980s and that union makes for an intriguing story in itself.

*A bull moose in its prime – the symbol of Algonquin.* Algonquin Park Museum Archives – MNR Staff.

Bubenik was fascinated with antlers. He spent a large part of his long career sketching and analyzing the geometry of the antlers of moose, elk and other cervids (deer) in an attempt to discern how form translated to function. Bubenik practically invented a language of his own to describe antlers in consistent, scientific terms. He wrote lengthy tomes full of terms like the "pseudocervicorn bauplan"[16] that only Bubenik himself understood.

Much of this was done for the fun of it but Bubenik knew that antlers have a serious side. Antlers are made of bone and are shed and regrown every year. The antlers of prime bull moose are huge, weighing up to 32 kilograms (70 pounds).[17] Growing these great masses of bone is a serious investment, a diversion of energy that moose, as we have seen, can ill-afford to lose. Nature does not countenance silliness; there has to be a reason for such extravagance. The payback, of course, is the increased likelihood of being selected as a mate – of passing on one's genes. Antlers are powerful releasers to a cow moose in heat. The very fact that growing antlers is so demanding signals that bulls possessing very large ones have that something extra – the sort of fellow you want fathering your children. Larger antlers generally belong to larger bulls, those which are almost always victors in the monumental one-on-one battles fought over cows. Oddly, though, it is not always the winner that ends up with the girl; there are well documented cases of cows, even entire harems, sticking with a losing bull.[18]

It was these sorts of innate mysteries of moose that Bubenik was driven to resolve. Antlers are powerful releasers to cows but, between bulls, they are as glowing beacons in fog to mariners. In open country, the bright reflection off the whitish palms of antlers can attract rival bulls from kilometres away.[19] Rutting bull moose have a compelling "search image" for antlers and the size and shape of antlers are powerful indicators of status. Thus, one- and two-year-old bulls, which have only spiked "tines" and no palm-like structure to their antlers, are completely ignored by older, prime bulls.

Antlers are the catalyst of elaborate rituals that have the effect of avoiding combat, injury and even death. For example, when two bulls come in contact, a ritualized sequence is enacted that often begins with the inferior bull "nosing" the proffered antler palms of the superior. If this works, and if the inferior bull keeps whining and signalling its submission, it can remain in the dominant bull's personal space. If not, the encounter will escalate to displays and "sparring" in which the posture and angle at which the antlers are presented are crucial to avoid conflict. Should the two bulls sense that they are evenly matched the engagement will escalate to fighting.

Cows fight too but, lacking antlers, they fight like horses – by rearing up and striking with their front hooves. These little scuffles can occur at any time of the year. In winter, when the rut is over, the testosterone level of bull moose drops to nearly undetectable levels. They become, in effect, eunuchs (testosterone is the hormone that produces male sex characteristics in all mammals). Interestingly, outside of the rut these "feminized" bull moose fight with their hooves as cows do; they "fight like girls."[20]

All of this was fairly well known before Tony Bubenik arrived in Algonquin Park, but Bubenik was consumed with his quest to understand every detail of moose behaviour. He felt that we could never properly manage moose populations without a full understanding of how their "pecking order" worked. To do this, he had to make detailed records of hundreds of moose interactions in the wild. This was a tall order, for even at places like Peter's Pond, which was a veritable moose jamboree, catching two moose in a social exchange was a chance event. To happen to get two animals of precisely the age, sex and social class he was interested in at the time was a near miracle.

Bubenik hit upon an ingenious solution; he would become a moose himself. He contrived a portable moose dummy to which could be affixed different heads and antlers and worn over his head (there were peepholes for seeing out). Sporting this preposterous contraption, Bubenik could act out any role he wished, changing his moose identity in seconds to test whatever experiment his real-life opponent called for. This superficially absurd idea was based on the notion that the antlers and the front end of the moose (the "head pole") were such powerful "releasers" that they were all that mattered. Rutting moose (bulls at least) would, in their state of ardour, ignore anything from the waist down and shoulders back.[21]

They did! Bubenik spent hundreds of hours under the dummy performing bizarre ballets with moose as equals in the arena of love and war. He essentially morphed into a moose, engaging the animals in intricate dialogues by varying his posture and movements in subtle and secret ways. You have to understand that the man *knew* moose; he understood, or would soon learn, every nuance of the body language of the magnificent beasts.

The spectacle of a man walking around with a moose head on could not have escaped the attention of the public nor, inevitably, the media. Bubenik enjoyed his greatest fame from a series of television news shorts and documentaries generated by the arresting images of a lone man challenging the great bull moose and the beguiling contradiction of a serious scientist doing such an apparently silly thing. He intended the material to be taken seriously: an attempt to teach the non-biologist something about the science behind his life's passion. It was, generally, and particularly the confrontations with large bulls (which showed real courage), but the images were pretty strange – strange enough that Bubenik was subject to some ridicule.

From his professional colleagues, though, Tony enjoyed nothing but respect, even reverence. Biologists lined up for a chance to work

*Tony Bubenik fashioning a dummy moose head for one of his experiments.*
Courtesy of Peter Smith.

with him. Peter Smith, from Ontario's wildlife division, was perhaps the most fortunate, forming a close working relationship with Bubenik that lasted a decade. Peter provides this unadorned but marvellous account of what it was like to work with Tony:

> "Dr. Tony Bubenik and I shared a passion, we both loved watching moose behaviour and I considered myself fortunate to have spent so many hours in the field with him in pursuit of our common interest …. Tony was a man of intense inner drives which were spurned on by an inexhaustible scientific curiosity. Tony used head poles (dummy heads) on a variety of deer family members in Europe and North America, i.e. red deer, caribou, moose. As part of a film sequence, Tony was required to construct a dummy female moose head.
>
> We obtained a road-killed cow moose from Algonquin and Tony and I prepared the specimen,

made plaster casts, and encased the cast in fibreglass. Tony just expected the work to be done, so I contributed several weekends plus evenings to the cause. It was my choice and I appreciated the chance to work with him. After a lot of work, filing and sanding excess fibreglass, the eyes and ears were inserted and Tony added the touch-up colour. Tony was meticulous in his detail. Things had to be just right. All the while, he wondered how would the cow moose respond to the dummy heads? What behaviour differences might be encountered? What if a calf was present? How close could you approach? Was visual stimulation sufficient? Were vocalizations needed?

We transported the head to Algonquin and tried to find a bull moose. After several days, we located a prospective suitor and I helped Tony on with the chest mounted dummy head. Before Tony left on his rendezvous with nature, he passed me his video camera. It was the first time I had seen, let alone used one. I was expected to record the event. The bull responded encouragingly to Tony's gestures (visual only), and after a brief minute or so, the encounter was broken off due in part to the marksman[22] who distracted the bull by continuously moving into better position. Tony returned to the truck with eyes a-gleam and sparkling and uttered a very common word for him 'fantastic.' It was 'fantastic.' It was the first time anyone, anywhere in the world had tried this. I had an opportunity to check the video later. It was awful quality, mostly out of focus and shaky, but I too reflected Tony's words. It was fantastic.

A few days after our first encounter, with a bull, Tony spotted a cow. He jumped from the truck and we hurriedly dressed him with the dummy cow head. As he departed, he said that I should don the bull head and follow him. I whispered back 'what should I do, call etc.' He implied that that was up to me. There was no script, no plan of action. It was all spur of the moment and ad lib."

Tony Bubenik passed away in 1995; Peter returns to study rutting moose in Algonquin Park to this day.

As it turns out, it was I who provided the cow moose head that became the model for the fibreglass replica Peter and Tony used that day. Peter had requested help from the Park in obtaining a cow's head and I got word out to the Park Wardens to let me know the next time there was a road kill. It being Algonquin, and the heyday of moose in the late '80s, I knew it wouldn't be long and we, sure enough, had a road-killed cow on Highway 60 within a few days.

The unfortunate animal had been struck hard and catapulted over a guardrail down a steep embankment. I arrived on the scene, ascertained that the moose was dead, and climbed gingerly down the rocky grade hefting a heavy axe. I started hewing the head off with great sweeping chops, but soon found myself in difficulty. Dismembering a moose (should you ever be called upon to do so) is no easy task; the neck is thick and sinewy, the spine like steel cord. Furthermore, my axe was dull. I laboured away in great frustration, cursing noisily until at length the last thread of skin was severed and the head departed the body. I gathered up the axe and head and started to clamber up the steep rocky slope when I saw, to my astonishment, that a large crowd had gathered. The group of several obviously city-bred families had, I suppose, stopped at the sight of an official vehicle at roadside. They had witnessed the entirety of my grisly performance. The children displayed emotions that varied from fascination to horror, the mothers a universal disgust and the dads a common stern concern. I reached the pavement gasping for breath with my gory trophy underarm and smattered from head to foot with blood and pieces of bone. Attempting to maintain a certain propriety, I nodded sagely, muttered something about "research," and forged resolutely through the parting bodies. I clambered into my truck and left before any questions could be asked, knowing that the truth was stranger than the grotesque scene that had just been enacted.

Smith relates another notable incident with rutting moose that involved Dan Strickland, a longtime naturalist in the Park (about whom we will hear much more shortly). It seems that Bubenik's arrival in the Park had inspired others to try their luck calling moose. One fine Saturday morning in late September Strickland was cruising a side road in a minivan when, rounding a corner, he encountered an enormous prime bull. The moose stepped off the road and Strickland tried an old hunter's trick to lure him out. The trick involves beating a moose antler against shrubbery to mimic the sound bull moose make when they "rub" and violently thrash bushes to attract attention. It worked. The great bull emerged immediately looking for his rival and Strickland, delighted at the response, "thrashed" some more. This continued, and the moose was lured closer and closer, until Strickland saw that the animal had developed an enormous erection and showed every sign of wanting to put it to use! Strickland retreated to the van, which he was expecting the bull to mount at any time, preoccupied with how he would explain the damage to the vehicle to his superiors. Fortunately, a Ford minivan, upon close inspection, is not sufficiently similar to a cow moose to "release" a sexual assault and the bull wandered off, no doubt bewildered by the whole affair.

Tony Bubenik and Peter Smith pursued their animated encounters with moose with great zeal for years and they learned a great deal.

They learned of the importance of face colour. Male and female moose have strikingly different face colouration. That of cows is light brown to an almost bright grey, while the face of bulls is a dark brown that, with rising testosterone levels in fall, turns nearly jet black. This basic and rather obvious feature of moose was overlooked until Bubenik (as he did with so many things) brought it to our attention. Bubenik showed that face colour is in itself a powerful releaser; contrasting shades of grey-black painted on dummies evoked completely different responses.[23]

They learned of the complex rituals of the rut. Fights without prior "rituals" like sparring are rare and ritual behaviour serves to allow bulls to test each other without resorting to fighting unless they are very evenly matched. Some of the "rules of engagement" are almost charmingly courtly. There is an unwritten rule, for example, that the victor of a fight does not maltreat the loser. This sort of chivalry apparently develops with maturity, as prime bulls will rarely strike out or try to gore a retreating rival whereas less mature bulls will, rather knavishly, do just that. Vanquished bulls, by the way, are not greatly daunted by defeat. They simply wander off to try their luck elsewhere, often on the perimeter of the victor's field.

They learned of the intricacies of posturing, the subtle body language that bulls use to avoid or start a fight. A bold frontal approach with head down to display the full span of antlers is a direct challenge and an indication of confidence. A lateral presentation, even very subtle dips to the side, signals submissiveness. In confrontations, smaller bulls almost always yielded to superior antlers by these gestures (to not do so, as Bubenik found using the dummy heads, risked provoking an immediate attack). With time, Bubenik learned all the tricks and was able to play massive bull moose like marionettes, turning their aggression on and off with cunning shifts of posture.[24]

This game-playing was, make no mistake, dangerous. The possibility of Bubenik provoking a charge was very real and even a single solid blow from a big bull could be fatal. As noted, Tony made sure that there was always a capable marksman with a rifle in the background in the event of things getting out of control. Mike Buss is an expert hunter and, knowing moose well himself, was often given the role of sentinel. Buss recalls a particularly dicey encounter with a large, aggressive bull during which the hostility of the animal seemed to mount uncontrollably despite Tony's efforts at appeasement by gesture and switching to smaller antlers. Bubenik was forced to retreat across the marsh until he was cornered at the observation tower at which point Buss was seriously considering shooting. Fortunately, Bubenik's store of false antlers was kept under the tower and, coolly switching to the biggest rack on hand, an exaggerated monstrosity of fibreglass, Bubenik forced the big bull to retreat. He had been, in a sense, released by a releaser.[25]

There is a limit to the tricks real moose can use to avoid fighting, and confrontations sometimes develop into combat. Serious fighting among moose, however, has long been assumed to be uncommon. Although there have been scattered reports of prolonged, nasty fights, some ending in death, biologists have assumed that the appeasement rituals generally worked. This may have to be re-examined in light of a recent study in Alaska in which it was found that upwards of ten per cent of bull moose have serious facial injuries after the rut and many were blind in one eye.[26] And then there is the most dramatic consequence of fighting – antler locking, in which, due to the convoluted geometry of points and palms, the antlers of two bulls may become inextricably locked together. This is a serious situation that, if not remedied, results in the death of both animals from exertion and stress. Although antler-locking of bull moose is allegedly not that uncommon in some areas, I am not aware of a case in Algonquin Park. All this aside, the really serious consequence of fighting is, as we have seen, the energy "wasted" by animals going into a long tough winter. But, if the result is the passing on of one's genes … well, what would Darwin have said?

Are moose "territorial"? The term, actually, is much misunderstood. All animals have a home range, but for a home range to be a territory it must, strictly speaking, be defended; it must be fought over. Moose illustrate this distinction perfectly. Moose have a home range but even big bulls are not usually hostile towards intruders; other moose are, by and large, free to roam around without fear of attack.[27] Even during the rut, when moose are aggressive, they are not really defending a specific piece of turf but rather a zone around themselves, a sort of personal space that moves with them (making them a sort of mobile territory).[28] Although territoriality is apparently not of great importance to moose, it is to many other species; the birds singing with such bucolic bliss in your backyard in spring are actually engaged in the deadly serious business of spacing themselves to avoid overcrowding.

Territoriality was a key to one of the most fascinating and insightful pieces of wildlife work ever done in Algonquin Park: a piece of sleuthing to rival Vasilauskas, Cole and Guyette, and perhaps even Anderson of brainworm fame. The detective in this case was Dan Strickland, longtime Chief Naturalist of Algonquin Park.

Dan Strickland is a brilliant dynamo of a man whose scholarship and potent writing and speaking skills have captivated two generations of Park visitors. An intensely intelligent and fundamentally decent fellow, Strickland has, at some point or other, studied and written about practically every item, event or trend in the natural history of Algonquin Park. His life's passion, though, has been the study of an inscrutable grey-black feathered enigma known as the grey jay.

Grey jays, often known as "whiskey-jacks," are another of the north country's icons. Slightly bigger than a robin, the grey jay wears a combination of black, white and shades of grey that, unexciting in themselves, combine to splendid effect. The family of birds known as the *Corvidae* (jays and their cousins, crows) are thought to be the most intelligent of birds and the grey jay is no exception. Grey jays are intensely inquisitive and seem to have an deep-seated drive to investigate anything that moves. They are curiously attracted to people and quick to learn and regularly visit bush camps to pilfer food, a habit for which they are famous and have acquired a reputation as "robber birds." It was this odd linkage with people that allowed Dan Strickland to unearth the fascinating strategies jays use to survive in the north woods.

Strickland studied grey jays first-hand for decades, beginning in the late '60s in La Verendrye Park in Quebec and from then in Algonquin to the present time. He focused his efforts on the nesting season, which is in March, and it is a common thing even today to see Strickland, dressed in a tattered old bush coat with his tousled grey-brown hair sticking out from under a worn toque, poking along the Park's southern thoroughfares in late winter in earnest pursuit of his beloved jays.

Perhaps because of their ghostly appearance and vaporous comings and goings, there is some strange folklore built around grey jays. One such myth, a legend still told by old woodsmen, is that "whiskey-jacks" don't reproduce as normal birds do but appear spontaneously as the beneficent act of some deity of the spruce woods. This foolishness must have arisen from the fact that grey jays nest in winter, much earlier than most birds, and tend to hide their nests in close to the trunks of thickly foliaged spruces and firs. Grey jay nests are therefore rarely seen and one can appreciate how the myth might have developed. Ironically, Strickland found that grey jay nests are exceptionally easy to find due to another of the birds idiosyncrasies: they readily take nest material, such as cotton balls, even proffered by hand, and thence lead the observer right to their nest site.

This was a great advantage, for the nests are usually located fairly close to ground and it was a simple thing for Strickland to approach with a stepladder and catch and band the young.[29] Sometimes, though, the nests were higher, in the insubstantial and unreachable tips of spruce trees. Such cases required ingenuity and Strickland had to jury-rig contraptions of lumber and ladders, sometimes dangling from great height and risk to reach the nest. That, however, was in his younger, foolish days.

By such means, Strickland was able to band essentially all of the grey jays in a broad zone straddling Highway 60. He used unique combinations of highly visible coloured bands to label each bird individually and, taking advantage of the jay's innate curiosity, was able to record thousands of encounters with the birds over a period of almost three decades. This massive record forms a sort of chronicle, a social history of the grey jay. It took dedication and many, many hours in the "bush" to acquire this profound understanding of the way of life of another species.

With time, an odd fact began to emerge. In fall, grey jays, when not observed alone, were almost always found in groups of three. This was not simply by chance, the number three was not, by statistical reckoning, likely to be the most common association.[30] Grey jays are monogamous, forming breeding pairs that are faithful until one or the other bird dies. The three-bird groups were always a pair with one of their young-of-the-year (usually a male) or a pair that had lost all of their own young in spring and "adopted" one (but always only one) juvenile from another brood. This was odd because there were plenty of young birds around; the adults normally rear three or sometimes four young from the nest and most of the young survive until fall.[31] Something was happening from late spring to fall to select one young bird from a brood to stay and cause the rest to leave. The ejected young were at a terrible disadvantage; almost 80 per cent died overwinter. Of the "stayers," which had the advantage of the parent's experience and protection, most survived. There was clearly a great advantage in being permitted, or having whatever chutzpah or luck it takes, to stay.

But why? And why only three birds? Intrigued, Strickland made special efforts to find out how and when the crucial split occurred. With

*A Grey Jay after the author's lunch!*
Courtesy of Norm Quinn.

great patience, because such things are a matter of diligence and luck, he managed to make enough observations of newly fledged young to determine that the "break-up" of the brood or family group begins in late May. Up until then grey jay families are at peace, but towards the end of May the young begin to squabble and one bird soon comes to dominate and expel all the others by early June. The ejected young are banished forever. Oddly, the parent birds are completely oblivious to all this, neither intervening nor assisting in the strife.

But again, why? Why do the adults sit idly by and allow all but one of the young that they have laboured so hard to raise (and, not incidentally, their genetic heritage) be sent to almost certain doom? And why always three birds, and why early June?

Strickland was convinced that the answer to these questions lay in another quirk of grey jay behaviour – food storage. Grey jays are one of the very few birds to have evolved the ability, like beaver, to store food for winter. Grey jays are busy all summer storing food, insects, mushrooms, etc., in any nook and cranny they can find, typically behind tree bark and under lichens. Storing food is necessary for the winter survival of some species in our northern forests. The north woods are a harsh, barren place in winter, and the very fact that grey jays store food suggests that there are not sufficient food resources available at that time. Grey jays also possess unusually large territories for such a small bird, about 70 hectares (180 acres), which also suggests that their resources are at a premium (i.e., they need more space to find enough food).[32] Strickland deduced that the adult pair may only be able to store enough food to support one additional bird through winter. Three birds seems to be all that a territory can support in winter and three is probably just the right number to help each other detect predators and share stored food. There seems to be a nearly absolute limit of three on grey jay winter families and a very harsh system had evolved to keep it that way.[33]

But the question remained, why are the subordinate birds ejected in spring? Why not allow the young at least the benefit of their parent's experience and protection through summer? The key to this question, Strickland deduced, was memory.

Grey jays have a phenomenal capacity for memory. The birds have to locate thousands of little stashes in winter by memorizing the locations from summer, not by just randomly searching for them. As such, their ability to orient in space and their "spatial" memory must be well beyond that of humans. Recall, now, that grey jays store their food during summer. It is thus vital for a young jay to be with his/her parents all summer so as to learn all of their hiding spots (and, incidentally, pilfer the tidbits and re-hide them, there being reason to believe that the act of hiding food, rather than just seeing it hidden, may be necessary to acquire the memory).[34] Strickland reasoned that it was necessary for the dominant juvenile to kick his brothers and sisters out *before* they had a chance to learn where all the food caches were.[35] The timing was essential; just having the competition gone wasn't enough because grey jay territories are so large that evicted young could easily sneak back unseen from the edges all winter and steal a meal. They could, that is, if they knew where the food was. In grey jays, nature has revealed again her impressive capacity to enforce cruel necessities.

But one question remains. Why do pairs that have lost their nests or young in spring accept an unrelated vagrant in fall? There can surely be no advantage to having to share their precious hoard of food and the transient birds do not, after all, carry the pair's genes. Dan Strickland never did figure this one out but suggested that it simply may not be worth the trouble (and precious energy) to drive the vagrants (which can be very persistent) away. There is also no doubt some advantage to have a third set of eyes on the lookout for predators.[36]

There is a striking parallel between Dan Strickland's grey jays and another member of the remarkable jay family – the Florida scrub jay. Here too, young jays stay with the adult pair rather than striking out on their own like most young birds, but for entirely different reasons. In the case of the scrub jays, it is the adults that benefit from the "stayers" as much as the young. Scrub jays inhabit small remnants of Florida oak scrub, a habitat that has dwindled in face of the blitzkrieg development of the State. Competition for space is intense and the jays maintain family ties to enhance their ability to defend their turf. Studies have shown that larger scrub jay families are able to hold larger territories and also raise more young.[37] The young jays that stay help feed their parent's next brood of young (highly unusual among birds) and families with lots of helpers do much better, raising an average of 2.4 young per year versus 1.6 for unassisted parents. But it is not so much the additional food but rather having more eyes around to detect predators, particularly hawks and snakes, that makes the difference. Scrub jays, it seems, are big enough to chase away at least some predators.

Strickland thinks that grey jays do not allow their young to help feed new nestlings because too much activity around the nest would *attract* predators. (Grey jays are apparently too small to drive predators away.) The main threat to young grey jays, incidentally, is red squirrels. Squirrels are not an animal one usually views as killers but they can actually be vicious predators of the nests of birds and even snowshoe hares. At any rate, the scrub jay family "system" tends to perpetuate itself because the territories of large and growing families eventually "bud" off new territories.[38] Like anything that meets nature's uncompromising litmus test, the passing on of one's genes, success breeds success and familial cooperation works for Florida scrub jays. (Scrub jays in the American Southwest, from which the Florida birds originated, have more than enough space to live and family groups dissolve after nesting, as is the case with nearly all birds.)

It is hard to imagine two places more different than the scrublands of central Florida and Algonquin Park, yet I am struck by the similarities between the two studies. One can picture the Florida researchers, binoculars dangling jauntily and notebooks in hand (but without the toques and coats) pursuing their colour-banded charges relentlessly –

and digesting thousands of observations into a penetrating analysis of the social life of birds – just as Dan Strickland did.

One would not expect the principals of this chapter, the moose and grey jay, to have much in common but they do and this came to light in the mid-'80s with work Ed Addison was doing on moose ticks.

Ed was observing a tick-infested moose when a pair of Dan Strickland's banded jays arrived on the scene and began foraging on the ground, attracted to bloody spots on the snow from the excreta of the ticks (yum!). Approaching closely, Ed saw that the grey jays were eating the grape-sized ticks. Intrigued, he approached Dan Strickland who, it turns out, had done a little research on the subject himself the previous year. The ever-inquisitive Strickland had placed an offering of bread (which was commonly used to attract jays) and eleven previously refrigerated ticks (I warned you about biologists) on a white sheet in front of a pair of jays. The jays ignored the bread (which was normally eaten eagerly) but ate the ticks with relish. One of the birds was seen to store a tick behind a curl of birch bark.[39]

Both biologists wondered how common this was because, if ticks were found to be a major source of food in late winter, it could have important bearing on the grey jay research. However, apart from some scattered and inconclusive observations, and one record from 1869, they could find no published accounts of grey jays eating moose ticks (although other birds, including scrub jays, were known to eat ticks off deer). But then in April of 1984, Strickland found a grey jay sitting on a nest of three young scattered amongst which were three toasty-warm moose ticks! At a loss to explain this, he speculated that the birds may place ticks that have just fallen off moose among their nestlings as miniature "hot water bottles" to keep the young warm when the parents are away.[40] (Another possibility is that ticks are a night-time food supplement for the female.) Regardless, the moose/tick/jay episodes are a reminder of the unlikely linkages that nature can conjure up. The liaison may be of some benefit to jays but, given the vast numbers of ticks in the bush, is probably of no great value to the wretched moose.

There is a disturbing epilogue to the grey jay story, because it now seems that Algonquin's jays may be in some trouble. Dan Strickland has noticed for some time that the birds are declining in the Park. What seems like the best territories are still occupied but marginal areas are being abandoned and the population declined by 40 to 50 per cent between 1970 and 2000. The reason for this is unclear but may have something to do with climate warming. The jays are breeding earlier (probably because of milder winters) but producing fewer young.[41] It may be that much of the food the birds store is perishing in the longer, warmer summers. With less food in winter, the birds are having a tougher time producing young. This is only a hypothesis but if it is true the future for Algonquin's grey jays may not be a bright one. They are at the extreme southern limit of their range.

Tony Bubenik and Dan Strickland are among the fortunate few people who have had the opportunity, and wisdom, to pursue a single problem in nature to its successful conclusion. But I have one more tale of determination – unfortunately the last but perhaps most fascinating serving of our Algonquin wildlife story. It is about turtles.

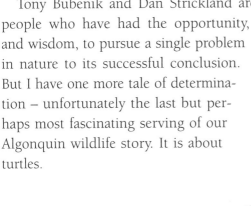

*Tony Bubenick at work.*

# 10 | Bet Hedging

> *The turtle lives twixt plated decks*
> *Which practically conceals its sex*
> *I think it is clever of the turtle*
> *In such a fix to be so fertile.*[1]
> — Odgen Nash, "Autre Bêtes, Autre Moeurs"

Snapping turtles are arguably the ugliest thing that the Creator ever made. (Whether the He or She actually made snapping turtles or not is another thing, but more on that in a minute.) Squat and massive, the olive-green brutes, which can weigh up to 18 kilograms (40 pounds) in Algonquin Park, have an exaggerated but not entirely unjustified reputation for nastiness. Even the scientific name of the snapping turtle, *Chelydra serpentina*, is evocative of evil.

The business end of a snapping turtle consists of powerful hooked jaws beneath tiny, malevolent black eyes all on a serpentine neck that, to the sorrow of the unwary, can be extended to an improbable length with astonishing speed. Perhaps the animal's most repugnant feature, however, is its stench. Snapping turtles are normally found encrusted with rotting algae and gross, fat, reddish-brown leeches, the combination of which, on close examination (not recommended), gives them an absolutely vile smell. All this makes for a most objectionable presentation and reputation.

Which is a great shame because snapping turtles are one of the most interesting of our native fauna and, unless one is quite stupid, entirely harmless. Snapping turtles are, in spite of suffering heavily at the hand of man, still a fairly common inhabitant of our bogs and swamps where they do no more harm than lolling about and chomping down on the occasional fish or frog. They actually consume a great deal of carrion as fisherman frequently find upon leaving a stringer of bass out for the night. Snapping turtles are, believe it or not, regularly taken by people for food, a real concern because of their exceedingly low reproductive rate (something of which we will hear much more about shortly). Their nemesis, though, is an unfortunate tendency to lay eggs in roadside gravel in spring, a habit that results in hundreds of turtles being run over (often deliberately) every year. A recent study in the Haliburton area southwest of Algonquin Park showed that as much as one third of the local population may be pancaked on highways every year, a rate of mortality that may not be sustainable.[2]

Although biting incidents are rare, they can be serious enough. I know of one case in which a young girl was bitten on the forearm and the turtle (having a brain the size of a pea) could not, in spite of every effort, be induced to let loose its vice-like grip. The animal had, eventually, to be decapitated and the young lady, with the turtle head still attached to her arm, taken to the local hospital where the grotesque accessory was surgically removed. It is this sort of thing, and their loathsome appearance, that has given snapping turtles their unfortunate reputation. (In truth, though, snapping turtles do not bite underwater, so swimmers are safe, and when they do bite on land it is really because they are terrified of us!)

To a small group of researchers at the University of Guelph, however, snapping turtles are a thing of beauty.

Ron Brooks, Professor of Zoology and the reigning czar of the Wildlife Station, has been conducting research on snapping turtles in Algonquin Park for nearly thirty years. His efforts, along with those of a dedicated corps of graduate students, amount to what is arguably the most important piece of scientific work ever done in Algonquin Park. And, as we shall see, among the most fascinating.

Brooks is a solidly built man of average height with a shock of curly grey hair and a mischievous twinkle in his eyes that exposes his inclination for spontaneous dry humour. Brook's thick-set build is the result of a youthful passion for boxing. I find this interesting for the only other boxer I have known is a doctor friend of mine who exhibited the same single-minded determination to get into medical school that Brooks has to get at the truth about snapping turtles. Boxing, in its savage simplicity, requires a discipline that has been likened to that demanded for pure pursuits of the intellect. It has been said recently that the skills required to defeat a boxer's defenses, the feinting, dodging and weaving, are about getting at the truth[3] – something, I suppose, like research. At any rate, Brook's office door in Guelph is adorned with scholarly articles that refute the commonly held belief that boxing diminishes one's mental capacities. This, in Brook's case at least, is most definitely true for he is the most intelligent man I have known.

Brooks' intellect and insight into the natural sciences are legendary. As an undergraduate at Guelph in the mid-'70s, I was one of a dozen or so students who would actually arrive early for his classes to contest for the best seats. We would crowd to the front of the lecture hall, in part because his delivery was rather subdued, but mainly so as not to miss any pearls of wisdom the man might drop. There was actually a sort of cult built around Professor Brooks, a small group of disciples who would follow after him in hopes of catching ear of a gem or two and then meet in cloistered session long afterwards. One detects a sense of this even today at the Wildlife Station where Brooks is usually seen in the company of a small entourage of admirers.

Ron Brooks' great passion is Evolution – the study of the convoluted strategies that have arisen through eons of time to allow species to reproduce successfully, avoid death, and thus up their odds of success. Biologists are baffled by the "debate" on evolution, seeing the evidence as overwhelming. Darwin, quite simply, was right – and right from the very start. To be sure, there is debate over some of the fine points of evolutionary change; there are, for example, puzzling "fits and starts" in the rate at which things have changed. But this is just mechanics, fine-tuning. No serious scientist any longer questions the so-called theory of evolution.[4] The "theory" is in fact the framework upon which all of the natural sciences, and indeed much of modern thought about life, death and society, are built.[5] Evolution *happened*; one only has to look at the exhaustive fossil records that trace the development in time of entire lineages, the horse,[6] for example, as proof.

One occasionally hears the argument that, in the case of humans at least, evolution is flawed because there are still "missing links" in the record. Actually, the common ancestry of apes and man has never seriously been questioned in science.[7] What is impressive, given the odds of finding four-million-year-old bones, is that so much material actually exists. Will archeologists have to unearth the remains of every proto-human that ever existed before the case is finally made? The argument is tantamount to finding a few bricks missing from the wall of a house and then declaring that, lacking bricks, it is no longer a house. But I digress; back to Brooks.

His specialty has been the study of herps (reptiles and amphibians) and the strategies these oft-neglected animals have evolved to maximize their production of young – and survive in a world that seems to have little use for them. There is probably no one in Canada right now who knows more about the biology and conservation of reptiles and amphibians, and particularly turtles, than Professor Brooks.

Most of Brook's work with herptilian sex has been done in a small lake (or more accurately a big pond) adjacent to the Wildlife Station. In this he picked the right place for, each spring with the warming and

*The business end of a Snapping Turtle.* Courtesy of Norm Quinn.

lengthening of the days, the pond becomes the scene of a veritable orgy of herptilian passion. Beaver ponds and swamps throughout Algonquin erupt with explosive mating activity in the month of June as turtles and frogs compete desperately to sow their seed in the narrow window of time allotted by the brief northern spring. Male snapping turtles run amok, copulating with anything that even vaguely resembles a female turtle including boulders, researchers' hats and, alas to say, other males. The other common turtle of the pond, the familiar and handsome little painted turtle, is no less crude. When rejected by a female, a male painted turtle will exhibit "spite" behaviour – churlishly nipping at her and inserting his head into her own head cavity to draw her out. Frogs, another of Brooks' study subjects, join in the raunchy sport. Several species of frog breed in June and the evening chorus of thousands of males rending variations of peeps, tocks, twangs and HARRUMPHS can be deafening.

Frogs are basic, dim-witted things and, it is said, demonstrate only three behavioural responses: if something is smaller than you, eat it; if it's the same size as you, screw it; and if it's bigger than you, run from it (sounds like some people I know). The bullfrogs at the Wildlife Station bear up this adage as, in the peak of their ardour, they have been known to mount and attempt to mate with anything their size including a researcher's proffered hand.

Reproduction, of course, is only one end of the population seesaw, the other being death or mortality. Any population of wild animals can be viewed as a leaky bucket beneath a flowing tap. The flow of water into the bucket (reproduction and immigration) is offset by the loss of water (mortality and emigration) through the rust-holes. The two are in constant opposition, a flux that sets the level of water in the bucket or, in our analogy, the number of animals in the population. Population biology is that simple, but the countless factors that influence the flow of water in and out of the bucket can, as we have seen, make it fiendishly complex. It has been Professor Brooks' mission in life to solve these mysteries.

Brooks has spent decades trying to understand life's ledger of birth and death for snapping turtles. Each spring, usually about the second week of June when the blackflies and mosquitoes are at their worst, the turtles of the station pond emerge to lay eggs on a few gravelly banks along the pond's shore.[8] Brooks and his students have been waiting for them since the early '70s. Once the turtles arrive, the researchers become furiously busy as they attempt to catch all of the ponderous reptiles in the very short time available. After nesting, each turtle is captured, taken to a nearby lab and (with due heed to the snapping jaws) weighed, measured and subjected to an array of probes and tests. Upon release, the turtles are free to go about their business until the following spring when the process is repeated.

Graham Nancekivell, a research assistant at the station, relates one of the frequent attempts to inject humour into the frenetic chaos of the team's labours in the nesting season. Some of the turtle nest sites are away from the station elsewhere in the Park and Brooks, perhaps just to get away for a while, but also because he has the memory of a grey jay and could best be trusted to locate and recall details of the nests, loved to check the "off station" sites himself. One day while Ron was out checking a nest, Graham found his Subaru parked in a sandy clearing near the Two Rivers campground. Nancekivell needed to talk to Brooks and decided to wait for him. As this took some time and, being idle and in a mischievous mood, Nancekivell decided to "fake" a nest in the sand. Being an expert himself, Nancekivell had no difficulty molding the sand with the characteristic depression and imprints of feet, claws and tail that a nesting turtle leaves. Upon returning, Brooks was about to leave with Nancekivell when the latter impishly pointed out this new "nest." Now Brooks prides himself on his observational skills and, annoyed that he had missed the nest, began to excavate it to tally and measure the eggs (which would then, of course, be reburied – under mesh cages to protect them from predators). Finding no eggs, the already annoyed Brooks expanded his slow, meticulous excavation all the while cursing his carelessness. Nancekivell, aware of the Professor's latent skills as a boxer, watched Brooks' antics with growing alarm, afraid the prank was going terribly awry. He eventually could take no more and announced that there "were no eggs," to which Brooks responded contemptuously that of course there were eggs and pressed on. Nancekivell finally 'fessed up and, to his great relief, Brooks took it well. Brooks, however, shortly had his vengeance with a counter-caper, the details of which I am not at liberty to reveal.

By such means the research team has, at any rate, been able to build an almost unprecedented chronicle of the inner workings of the station's snapping turtle population. Unlike most investigations of wildlife, the "demographics," or tally of birth and death, of Brooks' snapping turtles are easily measured. Each turtle is individually marked and the egg production of each female counted every year. Turtles can, furthermore, be aged (to a point) by counting growth lines on the "scutes" or bony plates that cover the shell (much as foresters count growth rings in trees). Brooks therefore has a record of the fortunes of the station pond's snapping turtles for nearly thirty years.

That record is, to say the least, a strange one. Most wildlife population show very rapid "turnover." Rabbits, for example, produce lots of young but suffer a high death rate, rarely living past three years of age. Elephants are a rare example of constancy, not maturing until eight to twelve years of age, producing only one calf every few years, and typically living to a ripe old age.[9] Brooks' snapping turtles carry this "actuarial conservatism" (the elephant's lifestyle) to an almost bizarre end; they just *are*. Snapping turtles, it seems, neither die nor are replaced. They just carry on forever.

Let me explain. Snapping turtle nests are extremely vulnerable to predators, mainly skunks, foxes and raccoons. The keen-nosed killers are able to detect the eggs and turtle hatchlings by scent, especially after a rainfall. Brooks found that due to this, and the frequent cool summers in which no eggs hatch, essentially zero baby snappers are ever produced. Even the few turtles that do hatch have basically no chance of survival; the research team has released over 6,000 hatchlings that they have incubated themselves and only one has survived to adulthood! So far, in fact, the odds of a hatchling living to 18 years of age (the age that the turtles mature) in the Algonquin study is exactly 0.000692 and less than one turtle a year is added to the adult population.[10] At the other end of the scale, the very few turtles that do survive essentially never die. The annual rate of death of the turtles at the station pond is less (probably much less) than one per cent, simple math telling us that, once mature, a snapping turtle can expect to live one hundred years.[11] The work of a Ph.D. student of Brooks', Marty Obbard, suggested that they may live considerably longer.[12] (No one knows for sure how old snapping turtles really are because at about forty-five years of age they stop growing and cease laying down growth lines. Counting the rings, an iffy process at best, then becomes useless.) Stranger, perhaps, is the observation that some of the adult turtles seemingly never put on nor take off *any* weight; Brooks has records of individual turtles whose weight has not deviated much more than a few ounces in more than a decade. Whether these individuals are sort of reptilian zombies – walking dead that don't function biologically (i.e. eat) – or not is unknown. Possibly they have very sensitive "weightostats," something many of us humans could use.

Brooks, at any rate, had documented a scientific enigma: an essentially stagnant population. He has published several top-drawer papers on the phenomenon and its implications for the conservation of turtles. After fifteen years it seemed the conclusion that the station's snapping turtle population was in a steady state, a sort of population limbo, was a safe bet.

Safe, that is, until, the spring of 1987.

In June of 1987, the turtles failed to show up. More accurately, only about half the customary number of females turned up to nest along the pond's sandy banks. Brooks assumed that the animals were just late until a search revealed turtle carcasses scattered all over the marsh. It was at first believed that the turtles had suffocated under the ice, because shallow northern ponds and lakes are often subject to "winterkill" from the depletion of oxygen. A close examination of the carcasses, however, suggested a different story. The turtles were nearly all mutilated in a similar manner; holes were gouged along the base of the shell near the legs and the insides ripped out.[13] Most also showed grooves on the plastron, or bottom shell, near the openings that appeared to have been made by the teeth of a predator. The killings continued for two more winters and by the time it was over Brooks had lost almost 70 per cent of his adult turtles. The source of the killings, however, remained a mystery.

In mid-winter of 1988-89 a radio-tagged turtle was found in the vicinity of an otter hole in the ice. Later that March, the same turtle was tracked 30 m away, near a second otter hole. That spring the animal was found again, alive, but just barely, his intestines and one limb chewed away. Turtles are prone to hibernate in groups in sheltered sites known as "hibernacula" and otter holes were found near several hibernacula that same winter. The issue was settled when Brooks took a cue from Bergerud's work with caribou calves (as we did with brook trout) and matched the gap of the canine teeth from an otter skull precisely to the gouges in the shells of the turtles.[14]

It seemed obvious that otters were the culprits, but the sudden onset of this killing was very odd because otters had always been present on the pond and yet had never before molested the turtles. Brooks could find only one other account of such behaviour – a case in Wisconsin in which otters were observed pulling snapping turtles onto the ice alongside a river to consume them.[15] He could only conclude that otters occasionally come by chance across a cluster of sleeping turtles and, being opportunists, take what they can. The incident is reminiscent of our own experience of otters ravaging spawning brook trout as discussed in an earlier chapter.

The whole business of the fortunes of Algonquin's snapping turtles is a striking illustration of the importance of doing long-term studies. Had Brooks walked away from the project in 1986, content that he had made his case, science would still believe that snapping turtles (in Parks at least) are models of stability. In fact, Brook's work shows how very vulnerable snapping turtles really are. Populations of animals that show very poor survival of young can ill-afford also to lose their breeding stock. High losses of both young and old are simply not sustainable. In the years following the catastrophe of 1987-89 Brook's snapping turtles did not "compensate" for their losses by increasing their production of young (as happens with some wildlife populations). Even if they had, the young would not have survived with all the predators about. Brooks employed sophisticated mathematical models to show that snapping turtles, like endangered sea turtles, cannot support heavy external killing.[16]

Nature imposes some difficult choices. Death being inevitable, there is only so much fuel in the tank. Alex Comfort has advanced the "rate of living" hypothesis, proposing, essentially, that a very heavy investment in reproduction early in life makes an early death inevitable.[17] Animals, ourselves included, have had to choose between producing young prodigiously and dying young, or (like us) having few young but over an extended period and living longer. These two basic strategies are known as "R-selected" (lots of young, short lifespan) and "K- selected" (few young, long-life). Brook trout are a good example of the former, producing thousands of eggs but living at most five to six years. Elephants and snapping turtles are an extreme example of the latter. Both strategies carry huge risks; trout, for example can saturate their world with eggs and young, but if conditions are unfavourable for only a few years, the population may not reproduce at all and die out. The approach taken by snapping turtles is less risky but because so few young are produced each year, it does not allow for rapid expansion into new areas when new opportunities arise.[18]

Snapping turtles, though, have it really tough because practically none of the very few young they produce survive. This forces them to adopt a strategy known as "bet hedging." In bet hedging, animals compensate for low and unpredictable survival of their young by spreading their reproductive years over a very long period (the K approach taken to an extreme).[19] Snapping turtles, by not putting all their eggs in one basket, are ensuring that at least once in a while, some will come through. It is a fundamentally cautious approach; they really are, in a sense, "hedging their bets." Professor Brooks' research has made snapping turtles the poster boy for bet hedging.

As if coping with raccoons, otters, road kill and having to hedge your bets was not enough, Algonquin's snapping turtles face another problem, a problem inherent to their basic biology. In 1971, scientists made the amazing discovery that the sex of European Pond turtle hatchlings can be determined by temperature.[20] Up to that point biologists had assumed that the expression of sex in animals was more or less straightforward and fixed in the XY chromosome pair. The idea that sex could be determined by anything other than genes was an outlandish concept. The work with pond turtles showed that at lower temperature of incubation, eggs are more likely to hatch as males; turn the thermostat right down and you get nothing but males! What causes this is not really clear but the phenomenon had been observed in several other species of turtles and Brooks confirmed it in Algonquin snapping turtles in 1994.

Algonquin is a cold place. Summers in the Park are short and there are frequent cool summers in which turtle eggs don't hatch at all. One does not have to be a population biologist to appreciate that the production of only male young is a losing venture. Algonquin's snapping turtles are at the extreme northern limit of their range. A prolonged series of cool summers, producing only males (or nothing at all) could jeopardize the turtle's survival in the Park and drive that range further south – another reason why our snapping turtles are bet hedgers. Fortunately for Algonquin's turtles, the trend seems to be the other way.

Peculiarities like sex-change in turtles have always appealed to Professor Brooks' fervent imagination. Perhaps nothing, however, will ever come out of the Wildlife Station to match for strangeness the curious story of the antler fly.

One fine mid-summer's day in the early '90s, Mike Tudor, station manager at the time, happened to notice tiny flies swarming on a moose antler that had been discarded near the cookhouse. Anywhere else such a thing would have gone unnoticed, but everyone at the station has his or her own sidelines and Tudor's was entomology. Upon close examination, he concluded that the flies were unusual, perhaps even a new species, and he mentioned the matter to Brooks. Intrigued, Brooks put a student, Russell Bonduriansky, onto the project.

Bonduriansky determined that the tiny (2.5 mm) flies which he dubbed *Protopiophila litigata*, were, sure enough, a new species, part of a family of minuscule flies that feed mostly on carrion. The discovery of a new species, even of a lowly fly, is always an event, but Bonduriansky's real achievement was in unearthing the astonishing life cycle of the antler fly.

Antler flies are extreme specialists, living their entire lives in and around discarded moose antlers. The high point comes on hot days in mid-summer when the antlers, erstwhile instruments of battle and romance, become themselves the arena for a bizarre Lilliputian scene of love and warfare. Bonduriansky deciphered the strange mating rituals of the antler fly through a combination of skill, patience and hard work. Improvising a miniature operating table on a microscope, he *painted* numbers on the backs of the tiny flies and then spent days hunched over an antler under the hot summer sun painstakingly recording their antics amidst a maddening horde of blackflies and mosquitoes.

At the peak of midday heat in early June, antler flies emerge from the surrounding leaf litter and swarm onto the upper surface of an antler where they engage in a furious communal orgy of fighting and copulation. The males face off in pairs spinning around furiously and "boxing" with their forelegs until one prevails, the winners being most likely to attract one of the dozens of amorous females cruising the arena. Paired off, the male penetrates the female with a formidable penis fully as long as his body and the happy couple retires to the shady underside of the antler to join hundreds of other pairs to mate in relative peace and discretion. The antler thus becomes a sort of miniature disco with hundreds of singles courting on the upper surface and couples mating below. Incredibly, if the antler is flipped over the flies immediately rearrange themselves accordingly. The act completed, the female returns to the upper surface and lays her eggs in cracks in the bone, (some of which are the result of moose combat). The eggs hatch into larvae which penetrate into the spongy interior of the antler to feed on rotting organic material. A year later, the larvae emerge, climb to the tip of the spike-like projections of the antler, bend into a crescent shape and spring off like a gymnast to land into the leaf litter and develop into adults – and continue the cycle.[21]

Having such a specialized life-cycle is risky; should antlers disappear from use so would the antler fly (antlers, though, have been around for some 40 million years).[22] Brooks and Bondurianksy puzzled over the direction that evolution is taking the antler fly. Bondurianksy had observed that male antler flies tend to select larger females as mates (presumably because they carry more eggs) while females tend to select, or are selected by, the larger males. Why, then, are antler flies not just getting bigger and bigger, that being where evolution seems to be driving them? The answer, deduced by Bondurianksy, is simple: the channels in the matrix of the bone in which the larva live are very narrow, setting an upper limit on the size of the fly (which is a comforting thought given the belligerent nature of the males!)[23]

As homely and ill-tempered as the snapping turtle is, there is another turtle in Algonquin Park that is its polar opposite. This is the charming little wood turtle, Algonquin's nearest thing to an endangered species. (Actually when you think of it, how fair is it for us humans to label animals as "good" or "bad"; I'm sure snapping turtles don't think of themselves as ugly.)

Wood turtles are strikingly handsome creatures. About the size of a baseball cap, the wood turtle has a carapace, or upper shell, that is a rich, striated composition of olive and yellow that appears to have been deliberately carved into an ornate pattern of grooves and flaring ridges – giving it the nickname "sculpted turtle." The exquisite shell, and bright orange neck and forelegs (that grade to scarlet in the eastern part of its range) combine to stunning effect. It is, however, the character of the turtle as much as its appearance that offers its great appeal. Wood turtles are remarkably responsive and seemingly fearless of people. Like any turtle, they withdraw into their shell upon being picked up but shortly emerge to (unlike their cantankerous snapping cousins) give their handler a trusting, inquisitive stare. Wood turtles are in fact remarkably tractable and responsive and make excellent pets.[24]

This is a great shame, for this affability has, in large part, led to the species' demise. Wood turtles were widely collected for the pet trade (and for food!) in the early part of this century and populations were decimated nearly everywhere. Incredibly, the illegal trade continues to this day – a wood turtle can fetch a couple of hundred dollars (U.S.) on the extensive black market in reptiles. The wood turtle has, as a result, disappeared from almost all of its former range – which was essentially the Northeastern U.S. and extreme southeast of Canada. It is considered to be "at risk" and legally protected in every state from Maine to Maryland and west to Michigan except, oddly, Pennsylvania where it seems to be holding its own. The wood turtle originally ranged throughout much of southern Ontario but now exists in only three or four restricted areas and the only viable population in the province may be in Algonquin Park. The species is critically at risk in Ontario and protected by law.

The wood turtle is the closest thing we have to a true tortoise, spending nearly all its time on land except for the brief mating period in spring. This makes it all the more vulnerable because it cannot normally retreat to water to escape predators. Worse yet, it is highly mobile, often ranging for miles and it is this wanderlust that exposes it to people, predators and, most tragically, cars. Wood turtles are regularly flattened on roads. The threat of this untimely demise is so great that Professor Brooks believes that the construction of roads through wood turtle habitat essentially dooms the animals.

Natural predators are also a big problem. Much of the area now occupied by wood turtles is developed as farmland and these landscapes of corn and soybeans support high populations of raccoons and skunks. Here again, being on land is a drawback, for coons and skunks love to attack wood turtles. They are not usually able to kill the turtles but, almost as if in spite, chew off their limbs. Studies have in fact shown that as many as 70 per cent of wood turtles have mutilated limbs or tails.[25] It is not unusual to see a wood turtle with one, two or even three limbs gnawed off gamely forging on through the woods. All in all, this array of hazards has the future looking rather bleak for the affable little wood turtle.

It had been known that wood turtles existed in Algonquin Park from scattered sightings that go as far back as the 1960s, but it was not until the late '80s that anyone took a serious look at them. (The locations of the turtles in the Park are, for obvious reasons, kept confidential.)

In the spring of 1987 we managed, with great difficulty, to find several wood turtles by searching for days in the thick brush in areas in which they had previously been reported. We affixed specially designed cylindrical radio transmitters to the rear of their shells. (Although I'm not certain, I believe we were the first to put radios on wood turtles.) The tiny transmitters had a broadcast range of about one kilometre and the results were impressive. Some of our turtles ranged all over the map; one female, for example, made a nearly straight cross-country jaunt of more than three kilometres to her nesting site each spring. This particular specimen moved so quickly that the first couple of years we lost her after nesting and it wasn't until the third year that, by searching far and wide, we picked up the faint signal that led us to her summer home. Our work (which was subsequently taken over by Professor Brooks) confirmed what people had

*Wood turtle – on the air!*

suspected for years: wood turtles are exceptionally mobile – a big factor in their downfall.[26]

There were some unusual spin-offs from the turtle telemetry. One day I received a phone call from an excited fisherman. The fellow, it seems, had been casting his bait into a reedy shallows the previous evening when he noticed one of the reeds moving erratically (wood turtles do, occasionally, go for a swim). Recognizing the silvery antenna for what it was, he became extremely apprehensive when it starting heading his way, apparently thinking that someone was hiding along shore and directing a remote-controlled bomb to his boat (the guy must have had some serious enemies). To his great relief, the antenna turned, headed to shore, and emerged atop a friendly and beautifully ornate little turtle.

Any discussion of wood turtles cannot overlook the bizarre phenomenon known as "stomping." Wood turtles are commonly observed alternately raising on each foreleg and then dropping to the ground in a staccato, drum-like fashion – something akin to a man doing one-handed push-ups and slamming his chest to the ground. As a rule, wild animals don't do things without good reason – natural selection tends to weed out frivolous behaviour – but the purpose of stomping remained a mystery for years.

Then a biologist in Pennsylvania proposed an explanation that, strange as it seems, has been pretty well confirmed. It appears that wood turtles stomp to imitate the sound of heavy rain, thereby luring earthworms to the surface for an easy meal. The worms surface in response to the drumlike pounding, presumably to escape drowning in their tunnels during downpours! Commercial worm-pickers use various noisemakers to exploit this and wood turtles have apparently also learned to use the worm's weakness to their advantage.[27]

This sort of quirky and charming behaviour is just one more reason to keep wood turtles around, but I'm greatly afraid for their future. Wood turtles are, like snappers, strongly K-selected, not maturing until at least eight years of age and producing small clutches of eggs. Also like snapping turtles, the vast majority of nests seem to fall prey to predators.

As if this was not enough, a new problem has emerged. A fly has recently been discovered that appears out of nowhere as wood turtle eggs hatch and lays its own eggs in the still attached yolk sac of the baby turtles. Where these flies come from and how they appear at just the right (or wrong) moment is a complete mystery – one, perhaps, for another of Brook's students. The fly eggs hatch quickly and, as little grubs, consume the soft tissues of the turtles and kill them. A very high proportion of newly hatched turtles fall victim to this fly – just another challenge for the beleaguered little wood turtle to face.[28]

It will take a determined effort to reverse the wood turtles slide to extinction. The whole problem emphasizes the importance of places like Algonquin Park where things like wood turtles have at least a fighting chance.

It should perhaps be surprising that an area as immense as Algonquin Park contains only one seriously at-risk species (the eastern hognose snake is also in serious trouble in Ontario but probably no longer exists in Algonquin). The lack of rare animals is actually rather easily explained. The majority of endangered species find themselves so because of the loss of their habitat. Algonquin's forests, although models of their type, are not particularly rare; there are maple-beech and pine forests throughout northeastern North America. The animals found in Algonquin are therefore, as a rule, quite common. We are, however, frequently visited by rare things; a pair of great grey owls, for example, nested in the Park a few years ago.

We are also, on occasion, visited by peregrine falcons, a once-endangered species with which the Park does have a history and another interesting story. We tend to hear only the doom and gloom about the environment and rarely the successes. Peregrine falcons are one of conservation's great success stories.

The cause of the demise of peregrine falcons is well-known: the widespread use of DDT contaminated the food chain which caused eggshells of some birds of prey to thin to the point of breaking. Peregrine populations plummeted to the point of extinction. In the early 1980s, with DDT gone, a concerted effort was made to bring peregrines back. One technique that was commonly employed is known as "hacking." Hacking involves raising young birds of prey on site in the wild in "hack" boxes. Hack boxes are cunningly designed to allow the birds room to move and grow and yet have little if any contact with their caretakers. The young falcons are fed dead birds, usually quail, remotely through a tube and, if everything is done properly, never see a human and are thus not conditioned or "imprinted" to our kind.

When the young falcons are ready to fly the box is opened, again remotely, and they are released to fend on the own. (One thing, incidentally, that always puzzled me about this is how the young falcons, which spend weeks in the hot, dry hack boxes, obtain water – it is not provided. I asked this of an expert and was told that birds, falcons at least, can "recycle" water from food through a basic biochemical pathway – Kreb's cycle – a bit of scientific trivia in the event that you find yourself on that millionaire show).

There were historic records of peregrine falcons nesting in Algonquin Park – on the great cliff on White Trout Lake for example – and Algonquin, being Algonquin, was naturally chosen as a place to "hack" young falcons. For almost a decade, biologists clambered up and down cliffs in the Park providing for young "hacked" peregrines, an activity that kept me in shape for several summers.

On July 8, 1985, the appointed day having arrived, I released four young peregrines at the imposing cliff on Whitefish Lake, visited today by a popular hiking trail. Had I checked the weather, I might have delayed a day or two, for the following day a mighty storm blew through. The winds were unusually intense and the front carried right through to the Maritimes and out to sea. To my dismay, only two of the falcons remained at Whitefish when the weather cleared. This did not necessarily implicate the storm, for newly released falcons are highly vulnerable to predators – great horned owls in particular – but I was convinced the weather was responsible and events proved me right.

About a month later we received incredible news. One of our missing falcons (which were, of course, banded), showing extraordinary good taste, had landed on a freighter carrying Scotch whiskey several hundred miles off the coast of Ireland. The bird was emaciated, exhausted and near death. The captain, an amateur naturalist, knew that he had something unusual and cared for the bird all the way across the Atlantic. Upon arriving in New York, he turned the bird over to the U.S. Fish and Wildlife Service which, through the band numbers, traced it to us. We briefly debated with the American authorities about returning the bird to Algonquin and starting over but determined that the best thing was to release it at a suitable site in New York State, which was done. I wondered long afterwards if the bird had passed by ships carrying freight less tempting than whiskey.

Of the approximately 70 peregrines "hacked" in Algonquin Park not one, to my knowledge, ever returned to nest (which, of course, is the whole point of the exercise). In retrospect, Algonquin may simply not be a very good place for peregrines. Peregrine falcons feed exclusively on other birds, catching them in mid-air, and are famous for the astounding speed and agility with which they do so. Peregrines therefore need areas rich in bird life, like marshes, to hunt over. Algonquin has substantial marshlands, and even some near our hack sites, but the Park's wetlands are mostly boggy and infertile and not particularly abundant with bird life. Our peregrines may have forsaken the Park for, quite literally, greener pastures. The Park, in fact, may never have been particularly good habitat for peregrines, the birds that did nest here being perhaps losers in squabbles over choicer territories.

# 11 | The Station

> *The trouble is, you think you have the time.*[1]
> — Buddha, circa 500 B.C.

In the spring of 1995, Steve Marshall, an entomologist from the University of Guelph, arrived at the shore of Scott Lake and started counting flies. Marshall's purpose, as part of a team of scientists studying the lakeshore/forest interface (as discussed earlier), was to document the biodiversity of the shoreline by focusing on just one group of insects. Insects, and flies in particular, are an incredibly diverse group and were the logical ones to choose. Surrounded by clouds of two of his subjects — blackflies and deer flies — Marshall artfully employed the tools of his trade, various clever traps and lures, to catch as many types of flies as he possibly could.

Back at the lab in Guelph, he and his assistants began the painstaking process of identifying the thousands of flies they had collected. Identifying insects, you should know, is an excruciatingly laborious task, a skill that only a handful of devotees like Marshall possess anymore (the biological world having gone totally "high tech"). Entomologists squint through microscopes for hours on end, counting and measuring the tiny "palps," "scutes," "stipes" and various other minuscule structures found in the nooks and crannies of the dried-up bodies of their subjects. The number, size and shape of these structures separate the various species. The work requires almost superhuman powers of concentration.

At first things went normally, as the common and expected flies were, one by one, identified and catalogued. Marshall's interest grew, though, when the list went over one hundred and the normally staid Department of Entomology began to stir as two and then three hundred species were found. The growing excitement turned to amazement as Marshall's count reached four and then five hundred different species. In the end, the team found that an incredible *six hundred* species of flies had been collected along a few metres and a heavy stone's throw in from the shoreline of Scott Lake. Understand that I am not talking here about six hundred flies, but six hundred *types* of flies.

Marshall's discovery stunned the small and close-knit community of researchers working in the Park. There was a general sense of wonder, even a sort of joy, that nature could provide such astonishing diversity of life in one place.[2] I recall wondering if the same list of flies could be found alongside any of the hundreds of similar, but heavily developed, lakes just outside the Park. I doubted it; at least some of the flies probably appreciate the peace and quiet.

Although I shared in the general delight over Marshall's work, there was a side of me that was troubled. The shoreline ecosystem work had been progressing nicely and some truly marvellous stuff, like the work with salamanders, had emerged. There was a sense among the scientists that a real understanding of the system was at hand. But just when you think you've come to understand things, someone comes along and finds six hundred species of flies.

Algonquin's Wildlife Station was established so that we could "understand things." The station started with humble beginnings in the mid-'40s, just as the guns in Europe were being silenced. Up to that point only two really significant pieces of wildlife work had been done in the Park – a study of snowshoe hare by D. MacLulich in 1937,[3] and another of grouse by the inestimable C.H.D. ("Doug") Clark (who eventually became the head of Ontario's wildlife management program).[4] Both studies were done in the remote centre of the Park, there being no base of operations in the developing southern

*Wildlife Research Station Main Lab – now the cookhouse, 1974.*
Algonquin Park Museum Archives #2159.

highway corridor. (Clark, incidentally, was trying to elucidate the mysterious ten-year cycle of ruffed grouse. He had a theory that it was somehow caused by parasites in the blood of the birds – an idea that has since been debunked.)

Fisheries research was already well underway in the Park – the Harkness laboratory on Lake Opeongo having started up in the late 1930s – and the wildlifers had some catching up to do. At first, work was done out of a cabin on Smoke Lake, but it soon became obvious that a more substantive facility was needed. Frank MacDougall, Park Superintendent at the time, asked Clark to set up a wildlife station and Clark selected a site near present-day Highway 60 that was reasonably accessible but at the same time secluded.

One of the first things that was done was to lay a "grid" of reference points with surveyor posts throughout the entire area. This gridwork, based on the forester's standard unit of measure, the venerable "chain," anchored in space much of the work that was to follow. Clark and his associates were, even then, thinking in the long term and tried to document every plant, stick, mouse and bug on the area with reference to their grid. The grid eventually became the basis of Bruce Falls' famous small-mammal trapline – the longest continuous tracking of a "mouse" population in the world – a study that carries on to this day.

All this was in 1945, and the early pioneers stayed that year in tent platforms. The cookhouse, site of the revered photo gallery, was built – complete with stone fireplace – in 1948. The first "lab" (if it justified the name) was built the following year and a crude bunkhouse in 1950. By the early '50s the place was fully (if rather tenuously) functional.

Much of the early work had more to do with forestry than wildlife (this was not unusual – as we have seen, the early pioneers of wildlife biology were nearly all foresters by training). Bruce Stephenson, a delightful gentleman now living in bucolic retirement on a farm near Gravenhurst,[5] was, for example, charged with trying to figure out how to regenerate yellow birch (a commercially valuable species) in face of the then-intense browsing by deer. (The yellow birch project, incidentally, led to the formation of a forest research station at Swan Lake elsewhere in the Park.)

In the mid-'50s the emphasis shifted to beaver. Beaver were (and still are) the mainstay of the wild fur industry and biologists were trying to determine how much of a "harvest" beaver populations could safely produce. Trappers were brought in to trap beaver all around the station and study the effects. This was the first serious use of the station in winter and things did not go particularly smoothly. The participants were, by necessity, crammed into the cookhouse, the only heated building on the grounds. "Heated" is used advisedly, for the cookhouse had only the stone fireplace and some nights were bitter cold. Snowfalls were intense in those years and one of the earliest "snow machines," an enormous Bombardier J5, had to be brought in to get the trappers out to the beaver ponds. Conditions were so difficult that normal trapping activity was quite impossible and, had it not been for the steady and relatively well-paying "Government work," the trappers would have quit.

Things, at any rate, progressed and the station went through a series of directors – and steady growth – into the '60s. The group of parasitologists celebrated in the second chapter of this book, Roy Anderson, Ed Addison and Sherwin Desser among others, arrived and became a dominant force on the scene for years. They had to share the stage, though, with the project that brought the station its greatest fame – Pimlott's wolf study. It was actually Rod Standfield, head of research in Ontario, who started the wolf study. Standfield, however,

*Bob Bateman boiling dirty clothes beside Lake Sasajewun. This is how we dealt with them in the early years at the Wildlife Research Station, 1949.*
Algonquin Park Museum Archives #6241 – Don Smith.

*J. Bruce Falls and Don Smith having lunch at the Wildlife Station in 1949.*
Algonquin Park Museum Archives #6241 – Don Smith.

had to move on and found Doug Pimlott, who was working (small world) with Beregerud's caribou in Newfoundland, to carry the flag. Wolves, as prominent as they are, were only one of dozens of animals that the station was developed to help "understand."

But it certainly worked. The volume of research that has emerged from Algonquin's unassuming Wildlife Station is, as we have seen, astounding. We have learned so very much in 50 years.

We've learned how a tiny roundworm can fell the mighty moose. We've learned how wolves ambush deer and why our hemlock is disappearing. We've learned that turtles hedge their bets and that otters are maybe not so nasty after all. We've learned of the sexual rituals of moose and dancing flies.

The range and scope of work done at the station is nothing short of incredible.

On the morning of May 15, 1962, two ornithologists, Jim Lowther and Dan Wood, paddled out onto the station pond to collect a loon. Having shot and retrieved the bird, they immediately noticed several blackflies crawling out from beneath the feathers of the head and neck. Flies continued to emerge from the loon as it was skinned back at the lab and by the end of the day the two biologists had collected several hundred. The blackflies were subsequently identified as the species *Simulium euryadminiculum*, a variety that was known to science but, until then, was thought to attack only ducks. It was determined later that the species, in fact, is attracted *only* to loons. (There are over thirty species of blackflies, some of which, like *S. euryadminiculum*, are incredibly fussy in selecting their victims.) Returning to the shore of the pond, Lowther and Wood noticed a swarm of blackflies around the boulder on which they had placed the bird. Then, to their growing amazement, they saw another horde of blackflies swarming around the soapy washwater they had cleaned the skin with and discarded near the lab.[6]

Intrigued, the two biologists finished preparing the loonskin (which included curing in moth balls) and, several weeks later, devised a little experiment. Having presumably obtained the appropriate permits (although recall that this was almost forty years ago), they collected and prepared skins of three other waterbirds: a grebe, a merganser and a gull. They then placed the three "control" skins with the loonskin on the shore of the pond and settled back to wait. Blackflies began to collect on the loonskin by the first evening. By the second evening 900 blackflies, all *S. euryadminiculum*, had been collected from the loon. No flies at all appeared on the merganser, grebe or gull. It seemed obvious that the fly was lured by some chemical attractant, a scent, possessed only by loons.

Lowther and Wood then shored up the deduction that blackflies "hunt" by scent (no small thing, given the universal plague they present to people) by referring to an observation that had been published in the late '50s. Female *S. euryadminiculum,* it seems, had been observed flying over a lake in northern Ontario in ever widening circles simultaneously dragging their feet lightly on the water. The flies were evidently "trolling" for loon scent![7] The two ornithologists, careful and proper scientists, made no sweeping conclusion from all this but suggested that blackflies, at least this type, must hunt by scent at least to the point of proximity to a loon whereupon (they suggested) they may cue in visually to the loon's dark green head and white "necklace."

The point of all this is not merely to illustrate the marvellous idiosyncrasies of blackflies but to point out that Lowther and Wood weren't even studying insects. They weren't, for that matter, even studying loons; Lowther was researching the behaviour of sparrows. The two had made a chance observation and then spent the large part of a summer building it into a significant discovery. There must be something in the air at the station that inspires creativity.

*Blackflies on a loon.*

The result of all this creativity has been, as we have seen, an awesome parade of discovery, perhaps unequalled from any one location on earth. This is something to be celebrated. And yet, in my gloomier moments, the very depth and complexity of all this knowledge causes me some despair. Wildlife populations are, almost by definition, horribly difficult things to study and understand. I'm not sure that we'll ever really understand the workings of single populations, let alone figure out ecosystems that present us with 600 species of flies along a bit of shoreline.

Ron Brooks thought he had snapping turtles figured out. After fifteen years of hard work he had nailed down the outlandish but very real workings of their lives and deaths. It was a scientific *tour de force*; no one had ever achieved such insight (at least with turtles). And then along came a few hungry otters to blow the thing to shreds. Roy Anderson spent years unraveling the puzzle of "moose sickness." Thanks to Roy's monumental accomplishment we know that brainworm kills moose. And yet, despite the determined efforts of several research teams, the real effects of brainworm on moose populations in the wild are (to put it kindly) poorly understood. Legions of biologists have studied predation but get ten biologists in a room and you'll get ten different opinions of its effects.

I suppose this is good – you know, healthy debate and all – and I'm not saying I don't like a challenge. I often wish, though, that you could put wildlife in a test tube and, like a chemist, get simple, clear-cut results. I sometimes, again in my bleaker moments, think the whole thing, the quest, may be hopeless, even a kind of sham. Maybe wildlife live where they do, and in the numbers they do, just because the many factors that provide for them are there and our attempts to read grand designs into things are hopeless.

I was at the Wildlife Station the other day. It was a chill, grey day in late November, the dreariest time of the year in Algonquin. Algonquin Park, despite its reputation for physical beauty, can at times be a rather bleak place. The Park's Highlands catch moist air coming off Lake Huron, and cool and condense it to cloud. When weather systems "stall" we can get days and days of overcast while the skies over the landscape surrounding the Park are clear. November is the worst for this bleakness; with its leaves down, the forest, its floor and the sky combine to create a disheartening tableau of shadowy greys and browns. I always look forward to the first snows.

On this particular day I had been driving near the station and turned in for a look. Most of the researchers are back in their lecture halls and labs at that point in the year, but occasionally there is someone hanging on to study something that happens only in late fall, or is there just for the peace and quiet of being there. I held hopes of some company and a chat but was to be disappointed; the station was empty.

I pulled up to the cookhouse and, the door being unlocked, went in for a look around. The room was damp and chilly, the wood stove being untended, and had the musty smell of old furniture and the recent wet autumn. The atmosphere was gloomy, almost funereal. I was struck by the contrast with the warm, boisterous summer noon-hour reveries I'm accustomed to experiencing. I was drawn, as always, to the bank of old photographs on the far wall and stared at them in the semi-darkness for a long moment. The faces of the station's faithful stared back at me across five decades and I tracked the change in them – the greying, the balding and the gradual stooping of shoulders. And I thought how sad it was that so many had gone.

And I thought of the old characters: the crusty but lovable Tony Bubenik and his tots of Grand Marnier, the affable Jim Fraser and his of Scotch, some just away, most gone forever. And I thought of what colourful characters they had been, the new generation being so politically correct, sober and sensible. I wondered if we would see their like again, or if their like were around – but it was I that had changed.

I thought of my own mortality, the aches I feel after a day in the field in places I didn't know I had. I thought of Henry Kujala, a Conservation Officer and friend who had volunteered to fill a vacant spot on a fisheries survey and died in the helicopter crash that followed. It was said afterwards that the best are often taken that way. I thought of the dozens of bright-eyed graduate students who had passed through the station and the long hours of hard, sometimes bitterly hard, work they put in – Russell Bonduriansky stooped for hours over his antlers in mid-summer heat, fisheries people out tagging lake trout at night in tiny work boats in the swells and bone-chilling cold of late October. I thought of their commitment, their unreserved buoyant optimism in the face of great challenge, poor pay, numbing work hours and, ultimately, poor prospects of employment.

And I thought again of the old guys: their dedication, their hopes and their defeats.

And I left the cookhouse in this grey mood, got in my truck and drove away. But then, as I turned onto the main highway, the sun broke through the clouds and hit my windshield and flooded the truck with light and warmth. And my mood changed in an instant. And I thought of the moose survey we would be doing this winter and whether the population had recovered from last winter's tick die-off, and if so why and if not, why not. And I thought of Stan Vasiliauskas and his surveys that show, with the moose population being down, a slight recovery of hemlock.[8] And I wondered if the respite will be enough. And I thought of last summer's cool weather and how it had wiped out the wood turtle hatch. And I wondered if that was really a problem and how many such years back to back might be the final blow. And I wondered how our otters were doing in Missouri and, for that matter, how they were doing in Algonquin Park, and whether we really needed to know that.

And I thought, what the hell, we know a whole lot more now than we did 50 years ago. And I thought that, if I have anything to say about it, the quest will continue.

# Appendix I

*Technical Expansion on Selected Topics*

The following sections are notes of a more technical nature that expand on some areas discussed in the main body of the book.

## A. Wolf Predation on Moose (Expansion from page 101, Chapter 4, Of Tooth and Claw)

Just as our impression of scavenging was changing so, slowly but surely, was thinking about the role of predation. Much of this was due to the growing body of research on wolves and ungulates (deer, moose, etc.) which made it increasingly obvious that wolves could, in some circumstances at least, very effectively depress prey populations.

Lloyd Keith, a professor of wildlife biology at the University of Wisconsin and perhaps the most prolific wildlife researcher of our time, was one of the first to acknowledge this fundamental change in thinking. Keith published a seminal paper on predation in the mid-'70s.[1] He believed that the newest research established clearly that predation can, and often does, depress moose and deer populations. Keith also remarked (taking care to pay respects to the great man first) that it was regrettable that Errington picked for study two species (muskrats and quail) that happened to be "self regulating" and thus prejudiced thinking on the subject for a generation.

Keith gave the nod to Bergerud's early work with caribou (which was just coming to light) but was most impressed with decisive events that had transpired on Isle Royale, a rocky, windswept island off the coast of Michigan's Upper Peninsula.

The island was devoid of both moose and wolves at the turn of the century, a natural laboratory or sort of gigantic Petri dish primed for culture. Moose arrived in 1908 and the population grew rapidly until 1935 when the animals depleted their food supply and starved *en masse*. This was repeated again in the late '40s. Wolves first arrived just after the second "crash" and by the mid-'60s had stabilized the wildly fluctuating moose population at about 1,000 animals. This was well below the number of moose that the Island's food supply could support. The chance sequence of events on this austere windswept island in Lake Superior was a model demonstration of the capacity of an unmolested wolf population to control events (wolf hunting is not permitted on the island).[2]

Keith then came to an inspired conclusion. The fact the moose (and deer) had been shown incapable, in the absence of predators, to limit their own numbers short of catastrophe was, he reasoned, *prima facie* evidence that under natural conditions predators (wolves) did so for them. His reasoning here was simple; if, prehistorically, moose had been subject to wild and catastrophic fluctuations (as they had on Isle Royale) natural selection should have developed means that would allow them to limit their own numbers before such disasters occurred. Keith knew that other wild animals had been shown to posses this ability, generally through territorial behaviour that distributed numbers across the landscape. The fact that moose and deer did not meant that they had never had to; wolves had always kept their numbers below starvation levels.[3]

One would think that this would have "iced" the issue, and indeed evidence accumulated through the '70s and '80s that wolves did control moose numbers. But nothing, it seems, is straightforward in wildlife research. Bergerud, for example, published a study in 1983 that suggested that wolves were limiting moose in northern Ontario. Sometimes publishing in science is like waving a red flag, and Bergerud's paper attracted a counter from biologists working in the area who held that the "problem" was food, weather and hunting, not wolves.

This quarrel developed into the classic "tit for tat" exchange that is occasionally played out in the scientific journals. These disputes carry a civil, almost gracious, tone in print, but often conceal bitter personal rancor. The tiff, in this case, was never really resolved.[4]

And so the debate about wolves and moose continues to this day but the balance of evidence is that, in anything approaching natural conditions, wolves will control moose numbers.

## B. Stocking and Genes *(Expansion from page 122, Chapter 6, Of Time and Trout)*

Finding evidence of interbreeding is one thing, but Danzmann and Ihssen were also looking to show how the incursion of hatchery genes might be affecting some vital biological character of Algonquin's wild trout. They collected eggs from brook trout from several lakes, determined the percentage of hatchery genes they carried, and then raised the fish separately under identical conditions in the laboratory and measured their growth. Since the laboratory environment and food ration were identical for each group of fish, any difference of growth among the lakes could only be attributed to the fish themselves – that is, to their genes. This would establish that at least one important trait (growth) differed, under genetic control, from lake to lake.

Growth was, as they suspected it would be, under strong genetic control. Trout from lakes with greater than 50 per cent hatchery genes were nearly 50 per cent larger at age two than predominantly wild fish.[1] The difference had to be hereditary and had to be at least partly due to the hatchery genes. Hatchery-reared fish tend to be large because of deliberate selection in the hatchery, generation after generation, for bigger and bigger fish. It was not surprising, then, that wild fish with a high component of hatchery "blood" should grow fast.

But now, you say, I've got you! – hatchery fish are supposed to be "bad," but grow bigger, which is "good."

An awkward point, I concede, but Algonquin's wild trout are what they are for a reason. It is very likely that the smaller size of Algonquin's native fish serves some purpose in survival. Maintaining a really large body size in the clear and clean (but infertile) lakes of the Park may be very difficult, particularly in face of intense competition for food with other fish like suckers, which can be extremely abundant in the lakes. Wild trout in Algonquin Park in fact rarely exceed five pounds. This is, make no mistake, a respectable trophy in anyone's books; however, there are probably good reasons that Algonquin's trout don't get much bigger. Trout in the Park must find it particularly difficult to get enough food to see them through the long winters when they do not feed (much) and can actually starve. If so, fish that were naturally "oversized" as a result of inbreeding with hatchery fish would be at a disadvantage. Another aspect of this is that simply having *bigger* fish might mean having *fewer* fish because the lakes can only support so much "poundage" of trout.

There is another downside to rapid growth. It was discovered in the lab that the hatchery-origin fish matured to spawn at a younger age. Again, this would, on the surface, seem to be an asset. As a rule, however, early maturity in fish means early death and Ihssen found, sure enough, that having hatchery genes shortened the life expectancy of the wild trout.[2] Wild fish therefore live longer than the "domestic" hatchery stocks and can spawn several times before dying. Most domestic brook trout released into the wild will spawn only once, if at all. By extending their reproductive life, Algonquin's trout are "putting their eggs in several baskets" and this may be another adaptation to the Park's harsh environment. Spawning populations are small and, as discussed earlier, a run of tough luck could eliminate stocks unable to bridge past a bad year to spawn successfully at least once.

## C. Carotenes (Expansion from page 132, Chapter 6, Of Time and Trout)

There is another aspect to the story of red-orange pigments, a topical and serious one, for carotenoids may play a major role in human health as well. There is growing interest in the role that beta carotene (vitamin A) and related compounds called retinoids play in regulating biological functions and preventing disease. Scientists are beginning to refer to the group as the body's "A team."[1] Retinoids, it seems, may regulate genes that control such fundamental functions as growth and development. It has long been known that Vitamin A deficiency results in (among other things) skin abnormalities resembling cancerous changes, and recent epidemiological studies suggest that beta carotene and retinol do indeed play a role in prevention of certain types of cancer (although this result is now being challenged). Beta carotene is also being recognized as one of a group of compounds that rid the body of potentially dangerous oxidizing agents[2] – and may even slow the aging process. Beta carotene has passed from the early adulation of the health food fraternity to the guarded acceptance of science as a beneficial dietary supplement and nutritionists now call for increased consumption of vegetables rich in Carotenes – carrots and broccoli for example.

## *D. Stress (Expansion from page 135, Chapter 7, Stress)*

We talk about stress a lot these days but what is happening here? What, exactly, is the stress response?

It is an amazing fact that exposure to almost any type of stress will cause an immediate spike in production of chemical transmitters in the brain that cause the rapid cascade of events known as the stress response. When the conscious inner brain of a man (or moose) is alerted to stress, impulses are sent immediately to the tiny pineal gland at the base of the brain. The pineal produces ACTH, a hormone that travels with great speed to the adrenal glands – inconspicuous pithy organs located above the kidneys (activation of the adrenals produce the familiar sharp "sinking" feeling of fear in the belly). Upon receiving ACTH, the adrenals, as their name might suggest, send adrenaline, a powerful hormone, throughout the body. Adrenaline prepares the body for imminent conflict; a series of physical reactions are invoked almost immediately. Blood pressure and pulse rate rise, and blood is shunted from the body's periphery to internal organs in case of injury. Adrenaline also mobilizes glucose (a type of sugar which is the body's short-term source of fuel) from the liver, to provide energy for the impending struggle.

Adrenaline gets all the publicity in our popular fixation with stress, but there is another group of hormones, the most important of which is Cortisol, that, over the long term, have more harmful effects. Adrenaline works by latching onto the surface of cells and activating a specific response (telling, for example, blood vessels to constrict – which raises blood pressure). Cortisol, a steroid, works by entering into the cell and manipulating its inner workings, rousing certain genes to action. The two hormones have some shared effects, but the effects of Cortisol tend to be longer lasting. Cortisol is a fundamentally "catabolic" hormone (that is, physically destructive, rather than "anabolic" which builds things up). If Cortisol is produced over a long period, it impairs and eventually shuts down certain body functions such as reproduction. Also, for reasons that are not fully understood, both Adrenaline and Cortisol, but particularly the later, suppress the immune system, causing sharp reductions in certain types of white blood cells that fight disease. Why this is so is not known, and something of an enigma. After all, under stress you need all the help you can get. It may have something to do with the main function of Cortisol, namely to make glucose available as energy at almost any cost. The sustained release of Cortisol can, at any rate, have devastating impacts on the health of both man and wildlife.[1]

Prolonged stress usually results in effects, such as chronic fatigue, that are slowly debilitating. The destruction of vital immune defenses can, however, be rapid and dramatic, particularly in wild animals that are captured and confined.

## E. Density Dependence (Expansion from page 139, Chapter 7, Stress)

J.J. Christian's ideas are part of a larger concept known as "density dependence." This idea here, quite simply, is that for any factor to control the number of animals in nature it must intensify as the population increases, eventually having a regulating effect. Perhaps the best illustration of the concept can be made with infectious disease. Diseases are often dormant when animals are scarce and contact between individuals is uncommon but become more prevalent as a population increases, eventually causing catastrophic epidemics. Most theories that have been put forward to explain the regulation of animal numbers, such as Christian's stress concept, are similarly "density dependent."[1] (Weather is an example of a force that is not; a hurricane will completely destroy a mangrove swamp and its inhabitants whether they are abundant or not.)

# Notes

## Frontispiece

1. O. Addison and E. Harwood, *Tom Thomson: The Algonquin Years* (Toronto: Ryerson Press, 1969) 126.

## Chapter 1: The Quest

1. Aldo Leopold, *A Sand County Almanac* (New York: Ballentyne, 1966) 22.
2. The artist is Paul Peel. Considered one of the best Canadian painters of his day, Peel was born in London, Ontario, in 1860 and studied in Philadelphia, London and Paris. He would frequently return to Toronto and London. He died prematurely in Paris of tuberculosis at age 32. The print is from "Canadian Masters Art Card Series, Stock No. 9692, Art Gallery of Ontario."
3. Aldo Leopold, *A Sand County Almanac*. Aldo Leopold is considered the "Father" of modern Wildlife Management. He was born in Burlington, Iowa, in 1887 and spent much of his early years as an outdoorsman and working for the U.S. Forest Service in the Southwest. Trained as a forester at Yale, Leopold taught biology at the University of Wisconsin from 1933 to 1938. He is famous for his passionate writings on conservation. Aldo Leopold died in 1948, fighting a forest fire.
4. H.G. Andrewartha, and L.C. Birch. *The Distribution and Abundance of Animals* (Chicago: University of Chicago Press, 1954).
5. Actually, the original Lascaux caves are now closed to the public for preservation reasons. However, a duplicate has been created for the public to visit.
6. Tim Byers, "Perspectives of Aboriginal Peoples on Wildlife Research," *Wildlife Society Bulletin*, 1999, 27 (3): 671-675.
7. Ron Tozer and Nancy Checko, "Algonquin Provincial Park Bibliography" (The Friends of Algonquin, *Technical Bulletin* 12, 1996).

## Chapter 2: Some Davids and a Goliath

1. Schopenhauer, "On Ethics" *Parenga and Paralipomena*, 1851, in John Gross, *The Oxford Book of Aphorisms* (Oxford: Oxford University Press, 1983) 5.
2. Norman Levine, *Nematode Parasites of Animals and Man* (Minneapolis: Burgess, 1968) 455.
3. Roy Anderson, "Neurologic Disease in Moose Infected Experimentally with Pneumostrongylus tenuis from White-Tailed Deer" (*Path. Vet.*, 1964) 289-322.
4. A.W. Cameron, *Report on Biological Investigations of Game and Fur-*

Bearing Animals in Nova Scotia (Nova Scotia Dept. of Lands and Forests, 1949) 5-28.
5. J.R.M. Innes, "Cerebrospinal nematodiasis, a Nervous Disorder Caused by Immature Nematodes (Setaria digitata)." (*North America Veterinarian*, 1952) 466-478.
6. Anderson, 15.
7. Ibid.
8. Albert Franzmann and Charles Schwartz, *Ecology and Management of the North American Moose* (Washington: Smithsonian Institute Press, 1998) 497.
9. Roy Anderson and Murray Lancester, "Infectious and Parasitic Diseases and Arthropod Pests of Moose in North America" (*Naturaliste Canadien*, 1974) 23-50.
10. Karen McCoy, and Tom Nudds, "An Examination of the Manipulation Hypothesis to Explain the Prevalence of *Parelaphostrongylus tenuis* in Gastropod Intermediate Host Populations," *Canadian Journal of Zoology*, 2000, 78: 294-299. McCoy and Nudds also make reference to the sheep flatworm study.
11. W.M. Samuel and D.A. Welch, "Winter Ticks on Moose and Other Ungulates: Factors Influencing their Population Size" (*Alces*, 1979) 303-348.
12. Edward Addison and Robert McLaughlin, "Growth and Development of the Winter Tick, *Dermacentor albipictus*, on Moose" (*Journal of Parasitology*, 1988) 670-678.
13. C.B. Blyth and R.J. Hudson, *A Plan for the Management of Vegetation and Ungulates, Elk Island National Park* (University of Alberta, 1987) 398.
14. Franzmann and Schwartz, 508.
15. S. Wickell, "Modulation of the Host Immune System by Ectoparasitic Arthropods," *Bioscience,* 1999, 49 (4): 311-320.
16. Mike Wilton and Dale Garner, "Preliminary Observations Regarding Mean April Temperatures as a Possible Predictor of Tick-Induced Hair Loss on Moose in South Central Ontario." (*Alces*, 1993) 197-200.
17. M.L. Drew and W.M. Samuel, Reproduction in the "Winter Tick, *Dermacentor albipictus*, Under Field Conditions in Alberta" (*Canadian Journal of Zoology*, 1986) 714-721.
18. Blyth and Hudson, 12.
19. Glenn DelGiudice *et al.*, "Trends of Winter Nutritional Restriction, Ticks, and Numbers of Moose on Isle Royale" (*Journal of Wildlife Management*, 1997) 895-903.
20. Dale Garner, *Population Ecology of Moose in Algonquin Provincial park, Ontario Canada* (Syracuse: State University of New York, Ph.D. Thesis, 1994) 81.

21. Tom Nudds "Retroductive Logic in Retrospect: The Ecological Effects of Meningeal Worms" (*Journal of Wildlife Management*, 1990) 396-402. The question of the real effects of brainworm on moose populations is still being debated.
22. Robert Warde and James McLeod, *The Zoology of Tapeworms* (New York: Hafner, 1968) 391-399.
23. Ibid.

## Chapter 3: Hemlock and History

1. Aldo Leopold, *A Sand County Almanac*, 265.
2. Cathy Keddy, *Forest History of Eastern Ontario* (Eastern Ontario Forest Group, 1993) 8.
3. Ibid, 9.
4. Ibid.
5. Henry Thoreau, *The Maine Woods* (New York: Thomas Cromwell, 1864) 198-199.
6. Craig Lorimer and Lee Frelich, "Natural Disturbance Regimes in Old Growth Hardwoods" (*Journal of Forestry*, 1994) 33-38.
7. Keddy, 27.
8. H. Lutz, "The Vegetation of Hearts Content, A Virgin Forest in Northwestern Pennsylvania" (*Ecology*, 1930: 1) 18.
9. Sam Graham, "Climax Forests of the Upper Peninsula of Michigan" (*Ecology*, 1941: 4) 360.
10. G. Nichols, "The Hemlock-White Pine-Northern Hardwood Region of Eastern North America" (*Ecology*, 1935: 3) 410.
11. Les Cwynar, "Recent history of Fire and Vegetation from laminated Sediment of Greenleaf Lake, Algonquin Park Ontario" (*Canadian Journal of Botany*, 1977) 15; J. McAndrews, "Late Quaternary Climate of Ontario: Temperature Trends from the Fossil Pollen Record" (Norwich: Geological Abstracts) 327.
12. Donald MacKay, *The Lumberjacks*. Toronto: Natural Heritage/Natural History Inc., 1998.
13. G. Whitney, "An Ecological History of the Great Lakes Forest of Michigan" (*Journal of Ecology*, Sept. 1987) 667.
14. Dan Strickland, *Trees of Algonquin Provincial Park* (Whitney: The Friends of Algonquin, 1996) 34.
15. Dan Strickland, *Algonquin Logging Museum: Logging History in Algonquin Provincial Park* (Whitney: The Friends of Algonquin, 2000) 1-4.
16. Mike Wilton, "How the Moose Came to Algonquin" (*Alces*, 1987) 23: 89-106.
17. Albert Franzmann and Charles Schwartz, *Ecology and management of the North American Moose* (Washington: Smithsonian Institute Press, 1998) 42.

18. Ibid.
19. Ibid, 47.
20. Randolph Peterson, *North American Moose* (Toronto: University of Toronto Press, 1955) 12.
21. James Burns, "Faunal Analysis of Two Sites in Algonquin Park, Ontario." (Unpublished manuscript, University of Toronto, 1972) 4; see Algonquin Park Archives.
22. Ibid, 9.
23. Ralph Bice, "Along the Trail, Caribou in our Part of the Province," *Almaguin News*, Oct. 4, 1972, 14.
24. Burns, 9.
25. Peterson, 16.
26. Ernest Thompson Seton, *Life Histories of Northern Animals* (New York: Scribner and Sons, 1909).
27. Burns, 8.
28. W. Wintemberg, "Lawson Prehistoric Site, Middlesex County, Ontario" (National Museums of Canada, *Bulletin 94*, 1939).
29. Roy Anderson, University of Guelph, personal communication in October 1999.
30. Wilton, 89.
31. Ibid, 90.
32. R. Runge and J. Theberge, "Algonquin: Decline of the Deer." (*Ontario Naturalist* 14: 20, 1974) 7-10.
33. Donald Robb, "A Beaver Census in Algonquin Provincial Park, 1939-40," (*Canadian Field Naturalist*, 1942) 86.
34. Wilton, 90-91.
35. C. Tufts, *Address to Forest/Fish and Wildlife Seminar* (Dorset, Ontario: Leslie Frost Natural Resources Centre, Feb. 1990).
36. Stan Vasiliauskas, "Interpretation of Age Structure Gaps in Hemlock Populations of Algonquin Park" (Unpublished Queen's University Ph.D. Thesis, 1995), iii.
37. K. Bennett, "Holocene History of Forest Trees in Southern Ontario" (*Canadian Journal of Botany*, 1986), 1798. Other authors refer to this ancient die-off of hemlock across its range and suggest that it may have been due to a pathogen, an agent that causes diseases such as a fungus or bacterium.
38. George Zug, *Herpetology: An Introductory Biology of Amphibians and Reptiles* (New York: Academic Press, 1993) 264.
39. Dave Cunnington and Ron Brooks, *The Role of Amphibians in Community Structure and Energy Transfer at the Aquatic-Terrestrial Interface* (Dept. of Zoology, University of Guelph, 1997) 18.
40. Richard Guyette and Bill Cole, "Age Characteristics of Coarse Woody Debris in a Lake Littoral Zone" (*Canadian Journal of Fisheries*

and Aquatic Sciences, 1999) 499.
41. R.A. Askins et al., Population Declines in Migratory Birds in Eastern North America in Current Ornithology (D.M. Power ed.) (New York: Plenum Press, 1990) 1-57.
42. Norman Duncan Martin, "An Analysis of Bird Populations in Relation to Plant Succession in Algonquin Park, Ontario" (Unpublished University of Illinois Ph.D. Thesis, 1956).
43. Andrew Smith, "An Analysis of the Changes in the Bird Communities of Four Habitats in Algonquin Park, Ontario from 1952-53 to 1995-96" (Unpublished University of Toronto M.Sc. Thesis, 1997) 82 pp.
44. C. Robbins et al., Population Trends and Management Opportunities for Neotropical Migrants, in: Status and Management of Neotropical Migrant Birds, D. Finch and P. Stangel, eds. (Fort Collins: U.S. Dept. of Agriculture, Gen. Tech. Rep. RM-229, 1993) 17-23. Scientists are at odds over the extent and causes of migrant bird declines.

## Chapter 4: Of Tooth and Claw

1. Paul Errington, Factors Limiting Higher Vertebrate Populations. in James A. Baily et al., eds. (Washington: The Wildlife Society, Readings in Wildlife Conservation, 1974) 167.
2. Errington, Of Predation and Life (Ames: Iowa State University Press, 1967) viii.
3. Ibid, vii.
4. Errington, Muskrat Populations (Ames: Iowa State University Press, 1963) 91. Much of what follows in this section is drawn from this book.
5. Ibid, 87.
6. Ibid, 88-89.
7. Ibid, 87.
8. Ibid, 101.
9. Information involving Ed Addison also appears in Chapters 2 and 10.
10. S. Kalter, "Cat Scratch Disease," in James Steele, ed. CRC Handbook Series in Zoonoses (Boca Raton: CRC Press, 1979) 425-431.
11. Errington, Factors Limiting Higher Vertebrate Populations, 173.
12. Ibid, 169-173.
13. Gordon Gullion, "Factors Influencing Ruffed Grouse Populations," Transactions of the 35th North American Wildlife Conference, 1970: 93-195.
14. Lowell Halls, White Tailed Deer: Ecology and Management (Harrisburg: Stackpole, 1984) 789.
15. Although fatal attacks are unknown on this continent, there have

| | |
|---|---|
| | been serious biting incidents in Canada in recent years, several of which were in Algonquin Park. Wolves in India kill several people every year. |
| 16. | Paul Wilson, et al., "DNA Profiles of the Eastern Canadian Wolf and the Red Wolf Provide Evidence for a Common Evolutionary History Independent of the Gray Wolf," *Canadian Journal of Zoology*. 1970 (78): 2156-2166. |
| 17. | See discussion on presence of moose and deer in Chapter 3. |
| 18. | The description of Douglas Pimlott at work on his final report comes from an interview with Bruce Stephenson on December 17, 2000. |
| 19. | Douglas Pimlott et al., "The Ecology of the Timber Wolf in Algonquin Provincial Park" (Ontario Dept. of Lands and Forests, *Fish and Wildlife Research Branch Report 87*, 1969) 11. Much of the following discussion of wolf research in Algonquin Park is based on this book. |
| 20. | Ibid, 11. |
| 21. | The source for these stories is Peter Smith, Ontario Ministry of Natural Resources, retired. |
| 22. | Pimlott et al., 32. |
| 23. | Tony Elders, "The End of Line 13," *ASKI*, 1993, Summer: 7-9. The incident has become something of a legend in wildlife management circles in Ontario. |
| 24. | Errington, *Muskrat Populations*, 93. |
| 25. | L. Ryell, *Evaluation of Pellet Group Surveys for Estimating Deer Populations in Michigan* (Michigan State University, Ph.D. Thesis, 1971) 237 pp. |
| 26. | Pimlott et al., 36. |
| 27. | Ibid, 35-38. |
| 28. | Ibid, 47. |
| 29. | The amount of fat in the marrow of the femur (leg) bones is used to measure the health of deer. |
| 30. | Pimlott et al., 40. |
| 31. | Tom Smith, "Hunting Kill and Utilization of a Caribou by a Single Gray Wolf," *Canadian Field Naturalist*, 1980, 94 (2): 175-177. |
| 32. | The legendary Ralph Bice became known as "Mr. Algonquin." Among his many achievements is his receiving the Order of Canada in 1984. Ralph Bice died in 1997 at the age of 97. For more on his life see Ralph Bice, *Along the Trail in Algonquin Park*. Toronto: Natural Heritage/Natural History Inc., 4th printing, 2001. |
| 33. | Pimlott et al., 62. |
| 34. | Lynn Rogers and David Mech, "Interactions of Wolves and Black Bears in Northeastern Minnesota," *Journal of Mammalogy*, 1981, 62(2): 434-436. |

35. D. Boyd and E. Heger, "Predation of a Denned Black Bear by a Grizzly Bear," *Canadian Field Naturalist*, 2000, 114(3): 507-508.
36. Douglas Pimlott, "Wolf Predation and Ungulate Populations," *American Zoologist*, 1967, 7: 267-278.
37. Dennis Voigt, "Changes in Summer Food Habits of Wolves in Central Ontario," *Journal of Wildlife Management*, 1976, 40(4): 663-668.
38. Alex Hall, "Ecology of Beaver and Selection of Prey by Wolves in Central Ontario" (Unpublished University of Toronto M.Sc. Thesis, 1971) 116 pp.
39. Arthur Bergerud, "Prey Switching in a Simple Ecosystem," *Scientific American*, 1983 (Dec.): 130-141. Much of the following narrative on caribou and lynx comes from this fascinating paper.
40. Ibid.
41. Ibid.
42. The introduction of the snowshoe hare was a spectacular success. Rural Newfoundlanders to this day rely heavily at times on snared "rabbits" for food; I have seen truckloads literally overflowing with hares on their way to roadside markets.
43. Bergerud, "Prey Switching in a Simple Ecosystem."
44. Ibid.
45. Ibid.
46. Ibid.
47. Vidar Marcstrom, *et al.*, "Demographic Response of Arctic Hares to Experimental Reductions of Red Foxes," *Canadian Journal of Zoology*, 1989, 67: 658-668.
48. Pimlott *et al.*, 81.
49. Russ Rutter, "A Sad Story," *The Raven*, 1971, 12 (1).

## Chapter 5: And Hares ... and Bears

1. Robert Blumenschine and John A. Cavallo, "Scavenging and Human Evolution," *Scientific American*, 1992 (Oct.): 90-96.
2. Paul Johnsgard, *Hawks, Eagles and Falcons of North America* (Washington: The Smithsonian Institute Press, 1990) 146-147.
3. Richard Conner, "Vocalizations of Common Ravens in Virginia," *The Condor*, 1985 (87): 379-388.
4. Mike Wilton, "Use of Time Lapse Photography to Determine Carcass Disposition," *Alces*, 1989 (25): 112-117.
5. William Gasaway, *et al.*, "Interrelationships of Wolves, Prey and Man in Interior Alaska, *Wildlife Monograph 36*, 1983. 50 pp.
6. Graham Forbes and John Theberge, "Importance of Scavenging on Moose by Wolves in Algonquin Park, Ontario," *Alces*, 1992 (28): 235-242. Much of the earlier discussion of the behaviour of wolves

around carcasses is drawn from this paper.
7. Alison Jolly, *Lucy's Legacy: Sex and Intelligence in Human Evolution* (Cambridge: Harvard University Press, 1999) 518 pp.
8. Blumenshine and Cavallo. Much of the discussion of scavenging and primitive man that follows comes from their article.
9. Ibid.
10. Ibid.
11. Ibid, 94.
12. Ibid.
13. Albert Franzmann and Charles Schwartz, *Ecology and Management of the North American Moose* (Washington: The Smithsonian Institute Press, 1998) 254-255.
14. Marty Blake, Ontario Parks, Personal Communication, June 2000. Author's Note: For a somewhat technical discussion of the "controlling" effects of wolf predation on moose see Appendix I.
15. Mike Wilton, "Black Bear Predation on Young Cervids: A Summary," *Alces*, 1983 (19): 136-147.
16. Mike Wilton, "Occurrence of Neonatal Cervids in the Spring Diet of Black Bear in South Central Ontario," *Alces*, 1984 (20): 95-105.
17. Robert Stewart, *et al.*, "The Impact of Black Bear Removal on Moose Calf Survival in East-Central Saskatchewan," *Alces*, 1985 (21): 403-418.
18. Franzmann and Schwartz, 262.
19. Dan Strickland, "What Can We Learn?" *The Raven*, 1992, 33 (1).
20. Strickland, "A Tragedy," *The Raven*, 1978, 19 (2).
21. Stephen Herrero, Address in Algonquin Park, May 19, 2000.
22. The best book on bear attacks, in the author's opinion, is Stephen Herrero, *Bear Attacks: Their Causes and Avoidance* (Piscataway, N.J.: Winchester Press, 1985).
23. Lloyd Keith, "Population Dynamics of Mammals," *International Congress of Game Biologists*, Stockholm, Sweden, 1973, 2-58.
24. Albert Lecount, "Characteristics of a Central Arizona Black Bear Population," *Journal of Wildlife Management*, 1982, 46: 861-868. See also his paper in the International Conference on Bear Research and Management, 1982.
25. Herrero, *Bear Attacks: Their Causes and Avoidance* (Piscataway: Winchester Press, 1985) 210-216.
26. Ibid.
27. Errington, *Of Predation and Life*, 241.
28. Ibid, 261.
29. The quote is believed to have been attributed to Lyndon B. Johnson, as read in an U.S. news magazine published during Johnson's political life.

## Chapter 6: Of Time and Trout

1. From "A Boy's Song" (1838) by Scottish poet James Hogg (1770-1835) in Elizabeth Knowles, *The Oxford Dictionary of Quotations*. Oxford: Oxford University Press, 1999.
2. J. McAndrews, "Late Quaternary Climate of Ontario: Temperature Trends from the pollen record" (Norwich: *Geological Abstracts*, 1981) 319-333.
3. Trout are actually one of the most ancient groups of fishes. They are perhaps the most highly prized of sport fish and have been transplanted all over the world, even to places like Tasmania, far from their natural range.
4. This effect accounts, in part, for the fact that fish are by far the most diverse of the vertebrates, having 20,000 species versus 4,000 for mammals, 8,600 for birds and 9,000 for reptiles and amphibians.
5. P. Lavin and J.D. McPhail, "The Evolution of Diversity in the Threespine Stickleback," *Canadian Journal of Zoology*, 1985 (63): 2632-2638.
6. Mary Keefe and Howard Winn, "Chemosensory Attraction to Home Stream Water and Conspecifics by Native Trout," *Salvelinus fontinalis*, from two Southern New England Streams, *Canadian Journal of Fisheries and Aquatic Sciences*, 1991 (48): 938-944.
7. Donald McIssac and Tomas Quinn, "Evidence for a Hereditary Component in Homing Behaviour of Chinook Salmon," *Canadian Journal of Fisheries and Aquatic Sciences*, 1988 (45): 2201-2205.
8. Fish have a well-developed sense of smell and orient to, among other things, fish dung, which may account for the fact that so many upriver brights stopped at the hatchery.
9. Peter Isshen, recently retired, was Chair of the Genetics Section of the American Fisheries Society.
10. Roy Danzmann and Peter Ihssen, "A Phylogenetic Survey of Brook Charr in Algonquin Park, Ontario, based on Mitochondrial DNA variation," *Molecular Ecology*, 1995, 4: 681-697.
11. Peter Ihssen, "Genetic Diversity of Algonquin Brook Trout Populations: Influence of Non-Native Genes," Ontario Ministry of Natural Resources, Unpublished Report.
12. Author's Note: For some more technical details of Danzmann and Ihssen's work see Appendix I.
13. Don McAllister, *et al.*, "Rare, Endangered and Extinct Fishes in Canada" (Ottawa: National Museums of Natural Sciences, *Syllogeus* #54, 1985) 44-48.
14. Jim Fraser, Ontario Ministry of Natural Resources, personal communication, Fall 1985.

15. J. Fraser, "An Attempt to Train Hatchery-Reared Brook Trout to Avoid Predation by the Common Loon," *Transactions of the American Fisheries Society*, 1974, 103(4): 815-818.
16. Jim Fraser, "Shoal Spawning of Brook Trout in a Precambrian Shield Lake," *Le Naturaliste Canadien*, 1985, 112(2): 163-174.
17. Ihssen, personal communication, 1990.
18. Ihssen, personal communication, 1986.
19. Scientists very recently took a new look at these "silver" trout using more sophisticated genetic techniques. Early results suggest that these fish may be more unique than we had originally thought.
20. The quote comes from an article "Frankenstein's Fish" by Ted Williams and John Huehnergarth in the late '80s in an unidentified popular fishing magazine.
21. Ole Torrissen, "Pigmentation of Salmonids: Effects of Carotenoids in Eggs and Start-Feeding Diet on Survival and Growth Rate," *Aquaculture*, 1984, 43: 185-193.
22. J. Craik, "Egg Quality and Egg Pigment Content in Salmonid Fishes" *Aquaculture*, 1985, 47: 61-68.
23. Ibid. The previous discussion of egg quality draws heavily on a review of the literature in this paper. For another, more technical slant on these pigments, see the section on carotenes in Appendix I.
24. Much of the discussion of colour in trout and trees is condensed from an article by the author in *Nature Canada* magazine in Fall of 1993.
25. W. Christie and H. Regier, "Temperature as a Major Factor Influencing Reproductive Success of Fish – Two Examples," *Rapp. P.-v. Reun. Contribution No. 71-4*, Research Branch, Ontario Department of Lands and Forests, 1973: 208-218.
26. Jim MacLean et al., "Temperature and Year-Class Strength in Smallmouth Bass," *Rapp. P-v. Reun. Cons, int. Explor Mer.* 1981, 178: 30-40.

## Chapter 7: Stress – Vulnerability Abounds

1. From Sophocles, "Fragment 58," Arcrisius 5th Century B.C. in John Gross, *The Oxford Book of Aphorisms*, 40.
2. Aaron Moen, "Effects of Disturbance by Snowmobiles on Heart Rate of Captive White-Tailed Deer," *New York Fish and Game Journal*, 1982, 29 (2): 176-183.
3. Claudia Wallis, "Stress! Seeking Cures for Modern Anxieties," *Time*, June 6, 1983.
4. Ibid.
5. The description of the stress response relayed here can be found in any general text on biology. For a more in-depth account see A. Guyton, *Textbook of Medical Physiology* (New York: Saunders, 1981). See also the *Globe and Mail*'s Web site article of Tuesday, Dec. 21,

1999, at www.globeandmail.ca. For additional information see also Appendix I in this book.
6. Marcia Baringa, "Where have All the Froggies Gone," *Science*, 1990 247: 1033-1034.
7. Joseph Pechmann *et al.*, "Declining Amphibian Populations: The Problem of Separating Human Impacts from Natural Fluctuations," *Science*, 1991, 253: 892-895.
8. John Christian, "Social Subordination, Population Density, and Mammalian Evolution," *Science*, 1970, 168: 84-90.
9. Ibid.
10. There are a number of studies that show the stress response invoked in wild animals as a result of increased abundance; see, for example, Bradley *et al.*, *General and Comparative Endocrinology*, 1980, 40: 188.
11. Madhusree Mukerjee, "The Population Slide," *Scientific American*, December 1998: 32-33. For more on J.J. Christian's thinking and his work on populations see Appendix I.
12. Murray Fowler, *Zoo and Wild Animal Medicine* (Philadelphia: W.B. Saunders, 1978) 789.
13. Ibid, 790.
14. R. George Brooks and Tomas Faley, *Exercise Physiology* (New York: John Wiley & Sons, 1984) 427-442.
15. E. Cross, "Arthritis Among Wolves," *The Canadian Field Naturalist*, January 1940, 2-4.
16. Charles Schwartz and Elizabeth Schwartz, *The Wild Mammals of Missouri* (Kansas City: The University of Missouri Press, 1959) 306-311.
17. Lloyd Keith, "Dynamics of Snowshoe Hare Populations," in H. Genoways, *Current Mammalogy* (New York: Plenum Press, 1990) 119-195.
18. Rudy Boonstra *et al.*, "The Impact of Predator-Induced Stress on the Snowshoe Hare Cycle," *Ecological Monographs*, 1998, 79 (5): 371-394.
19. Scott Creel *et al.*, "Radiocollaring and Stress Hormones in African Wild Dogs," *Conservation Biology*, 1997, 11 (2): 544-548.

## Chapter 8: The Twig Eaters

1. From Democritus of Abdera, 5th to 4th Century B.C., in John Gross, *The Oxford Book of Aphorisms*, 35.
2. Gardiner Bump *et al.*, *Ruffed Grouse: Life History, Propagation, Management* (Albany: The New York State Conservation Dept., 1947) 198.
3. Aldo Leopold, *A Sand County Almanac* (New York: Ballantyne, 1966) 230.

4. Douglas Morse, *American Warblers* (Cambridge: Harvard University Press, 1989) 210-211.
5. Joseph Sullivan and Susan Payne, "Aspects of History and Nestling Mortality at a Great Blue Heron Colony, Quetico Provincial Park, Ontario," *The Canadian Field Naturalist*, 1988, 102 (2): 237-241.
6. Lawrence Ellison, "Seasonal Foods and Chemical Analysis of Winter Diet of Alaskan Spruce Grouse," *Journal of Wildlife Management*, 1966, 30 (4): 729-735.
7. P. McDonald *et al.*, *Animal Nutrition* (Edinburgh: Oliver and Boyd, 1966) 454.
8. A. Sinclair *et al.*, "Diet Quality and Food Limitation in Herbivores: The Case of the Snowshoe Hare," *Canadian Journal of Zoology*, 1982, 60: 889-897.
9. Steve Gurchinhoff and William Robinson, "Chemical Characteristics of Jackpine Needles Selected by Feeding Spruce Grouse," *Journal of Wildlife Management*, 197, 36 (1): 80-87.
10. Harry Lumsden, Ontario Ministry of Natural Resources, (ret.), personal communication, July 1976.
11. B. Pendergast, and D. Boag, "Nutritional Aspects of the Diet of Spruce Grouse in Central Alberta," *The Condor*, 1971, 73 (4): 427-443.
12. David Fraser, and E. Reardon, "Moose-Vehicle Accidents in Ontario: Relation to Highway Salt," *Wildlife Society Bulletin*, 1962, 10: 262-265.
13. Albert Franzmann and Charles Schwartz, *Ecology and Management of the North American Moose* (Washington: Smithsonian Institute Press, 1998). These figures are an extrapolation of data presented in Chapter 12 of Franzmann and Schwartz's book.
14. Bill Pruitt, "Snow as a Factor in the Winter Ecology of the Barren-Ground Caribou, *Arctic*, 1959, 12 (3): 159-179. Pruitt, known as "the snowman," spent a lifetime studying snow and how wildlife live and die in it.
15. L. Renecker, and R. Hudson, "Seasonal Energy Expenditure and Thermoregulatory Response of Moose," *Journal of Wildlife Management*, 1986, 64: 322-327.
16. Ibid.
17. J. Kelsall and E. Telfer, "Biogeography of Moose With Particular Reference to Western North America," *Le Naturaliste Canadien*, 1974, 101: 117-130.
18. Franzmann and Schwartz, *Ecology and Management of the North American Moose*. A good overview of the biology of rumination is given in Chapter 13.
19. Ibid, 407.
20. Ibid, 403.
21. Again, for an excellent treatment of this subject, see Franzmann and

Schwartz, 1998, Chapter 13.
22. Ibid, 433.
23. Charles Schwartz et al., "Seasonal Dynamics of Food Intake in Moose," *Alces*, 1984, 20: 223-244.
24. Anthony Bubenik, "North American Moose Management in Light of European Experiences," *Proceedings of the North American Moose Conference and Workshop*, 1972, 8: 279-295.
25. Lowell Halls and Cindy House, *White-Tailed Deer: Ecology and Management* (Harrisburg: Stackpole, 1984) 115.
26. Franzmann and Schwartz, *Ecology and Management of the North American Moose*, 98.
27. Randolph Peterson, "A Review of the General Life History of Moose," *Le Naturaliste Canadien*, 1974, 101: 9-21.
28. Norman Quinn and Robert Aho, "Whole Weights of Moose From Algonquin Park, Ontario, Canada," *Alces*, 1989, 25: 48-51.
29. P. Jordan, "Aquatic Foraging and the Sodium Ecology of Moose: A Review," *Swedish Wildlife Research*, 1987, 1: 119-137.
30. Randolph Peterson, *North American Moose* (Toronto: University of Toronto Press, 1974) 280 pp.
31. R. Edwards, "Damned Waters on a Moose Range," *Murrelet*, 1957, 38: 1-3.
32. W. Regelin, *Deer Forage Quality in Relation to Logging in the Spruce-Fir and Lodgepole Pine Type in Colorado* (Colorado State University, Fort Collins, M.Sc. Thesis, 1971) 50 pp.
33. Franzmann and Schwartz, 423.
34. W. Regelin, et al., "Energy Costs of Standing in Adult Moose," *Alces*, 1986, 22: 83-90.
35. Franzmann and Schwatrz, 471-473.
36. R. Crawford, *Studies in Plant Survival* (Oxford: Blackwell Scientific, 1988) 227.
37. D. Ullrey et al., "Digestibility and Estimated Metabolizability of Aspen Browse for White-Tailed Deer," *Journal of Wildlife Management*, 1972, 36 (3): 885-891.
38. Crawford, *Studies in Plant Survival*, 227.
39. Ibid.
40. A. Starker Leopold et al., "Phytoestrogens: Adverse Effects on Californian Quail," *Science*, 1976, 191: 98-99.
41. Ian Baldwin and Jack Schultz, "Rapid Change in Tree Leaf Chemistry Induced by Damage: Evidence for Communication Between Plants," *Science*, 1983, 221: 277-279.
42. Irene Shonle and Joy Bergelson, "Interplant Communication Revisited," *Ecology*, 1995, 76 (8): 2660-2663.
43. N. Thompson Hobbs, "Modification of Ecosystems by Ungulates,"

44. John Fryxell and Charles Doucet, "Diet Choice and the Functional Response of Beavers," *Ecology*, 1993, 74 (5): 1297-1306.
45. Donald Robb, "A Beaver Census in Algonquin Provincial Park, 1939-40," *The Canadian Field Naturalist*, 1942: 86-90.
46. Ernest Thompson Seton, *Life Histories of Northern Animals* (New York: Scribner and Sons, 1909) Vol. 2: 1067.
47. Ibid, 1054-1087.
48. M. Gates and Bruce Falls, "Twenty Eight Years of Fluctuations in *Peromyscus* Populations in Algonquin Park" (University of Toronto, Dept. of Zoology, Unpublished Report, 1979) (see Algonquin Park Museum Library #3105). This is just one of a number of reports on this study which is said to be the longest continuous tracking of any small mammal population in the world and continues to this day.
49. Reto Zach and J. Bruce Falls, "Ovenbird Hunting Behaviour in a Patchy Environment: An Experimental Study," *Canadian Journal of Zoology*, 1976, 54: 1863-1879.
50. Ibid.
51. Gunnar Svardson, "Competition and Habitat Selection in Birds," *Oikos*, 1949: 157-174.

## Chapter 9: Moose Days and Jays

1. D.H. Lawrence, "Self Pity," in *The Reader's Digest Treasury of Modern Quotations* (New York: Reader's Digest Press, 1975) 758.
2. Stephen Herrero, Address in Algonquin Park, May 19, 2000.
3. Eugene Linden, "Can Animals Think?," *Time*, Sept. 6, 1999: 48-54.
4. Much of the anecdotal information related here comes from Mike Buss, a biologist who worked with Tony Bubenik and who is now retired from government service.
5. Anthony Bubenik, *et al.*, "The Significance of Hooves in Moose Management: A Preliminary Study," *Proceedings of the North American Moose Conference and Workshop*, 1978, 14: 209-236.
6. Bubenik, "North American Moose Management in Light of European Experiences," *Proceedings of the North American Moose Conference and Workshop*, 1972, 8: 279-295.
7. Bubenik *et al.*, 1978, 212.
8. This story is also from Mike Buss.
9. H. "Tim" Timmerman, *Morphology and Anatomy of the Moose Bell and its Possible Functions* (Lakehead University, M.Sc. Thesis, 1979) 90 pp.
10. For a full discussion of this difficult concept, see Albert Franzmann and Charles Schwartz, *Ecology and Management of the North American Moose* (Washington, Smithsonian Institute Press, 1998) 174-175.

11. Bubenik, "Behaviour of Moose of North America," *Swedish Wildlife Research*, 1987, Suppl 1: 333-365.
12. Much of the description of the moose rut provided here is inspired by Bubenik, 1987. For a more in-depth treatment the reader should refer to that paper.
13. Again, for more detail see Bubenik, 1987.
14. Ibid, 339.
15. Ibid, 360.
16. Franzmann and Schwartz, 104.
17. Jack Scott, *Moose* (New York: G.P. Putnam's Sons, 1981) 22.
18. Bubenik, 1987, 345.
19. W. Knowles, *An Ethnological Analysis of the Use of Antlers by Rutting Bull Moose* (University of Alaska, Fairbanks, M.Sc. Thesis, 1984) 93 pp.
20. Valerius Geist, "On the Behaviour of North American Moose in British Columbia," *Evolution*, 1963, 20: 377-416.
21. Bubenik 1987, 345.
22. The research was inherently dangerous and required the presence of a skilled hunter, just in case.
23. Bubenik, 1987, 342.
24. Again, see Bubenik, 1987, for a full treatment.
25. The story is from Mike Buss.
26. Franzmann and Schwartz, 205
27. Ibid, 189.
28. Ibid, 324.
29. It is an "old wives' tale" that birds abandon their eggs or young if touched by human hands.
30. Dan Strickland, "Juvenile Dispersal in Gray Jays: Dominant Member Expels Siblings from Natal Territory," *Canadian Journal of Zoology*, 1991, 69: 2935-2945.
31. Ibid.
32. Ibid.
33. Ibid.
34. K. Bunch, and D. Tomback, "Bolus Recovery by Gray Jays: An Experimental Analysis," *Animal Behaviour*, 1986, 34: 754-762.
35. Strickland, 2943.
36. Ibid. This paper is a testament to Strickland's perseverance and incisive thinking.
37. John Fitzpatrick and Glen Woolfenden, "The Helpful Shall Inherit the Scrub," *Natural History*, 1984 (5): 55-63.
38. Ibid.
39. Ed Addison, *et al.*, "Gray Jays and Common Ravens as Predators of Winter Ticks," *The Canadian Field Naturalist*, 1989, 103 (3): 406-408.

40. Ibid.
41. Dan Strickland and Thomas A. Waite, "Decline of the Gray Jay in Algonquin Park," *PRFO Occasional Paper Series*, 2001, 1: 11

## Chapter 10: Bet Hedging

1. Ogden Nash (1902-71), "Autre Bêtes, Autre Moeurs" from Elizabeth Knowles, *The Oxford Dictionary of Quotations* (Oxford: Oxford University Press, 1999) 539.
2. Tim Haxton, "Road Mortality of Snapping Turtles in Central Ontario During the Nesting Period," *The Canadian Field Naturalist*, 2000, 114 (1): 106-110.
3. Michael Winter, "Pugilists of the Page," *National Post*, Sat. June 18, 2000.
4. Ernst Mayr, "Darwin's Influence on Modern Thought," *Scientific American*, July 2000: 79-83.
5. Ibid.
6. Ruth Moore, *Evolution* (New York: Time Incorporated, 1962) 112-113.
7. Mayr, 282.
8. Turtles need sandy or gravelly spots with a warm southern exposure to dig nests and lay their eggs.
9. E. Walker, *Mammals of the World* (Baltimore: John Hopkins Press, 1968) 1319-1324.
10. Ronald Brooks, et al., "Effects of a Sudden Increase in Natural Mortality of Adults on a Population of the Common Snapping Turtle," *Canadian Journal of Zoology*, 1991, 69: 1314-1320.
11. David Galbraith and Ronald Brooks, "Survivorship of Adult Females in a Northern Population of Common Snapping Turtles," *Canadian Journal of Zoology*, 1987, 65: 1581-1586.
12. Martin Obbard, *Population Ecology of the Common Snapping Turtle in North-Central Ontario* (University of Guelph Ph.D. Thesis, 1983).
13. Brooks *et al.*
14. Ibid.
15. Ibid.
16. David Cunnington and Ronald Brooks, "Bet Hedging Theory and Eigenelasticity: A Comparison of the Life Histories of Loggerhead Sea Turtles and Snapping Turtles," *Canadian Journal of Zoology*, 1996, 74: 291-296.
17. For a fascinating treatment of aging and why it is biologically inevitable see A. Comfort, *The Biology of Senescence*. New York: Elsiver, 1979.
18. Author's note: For another fascinating view of why aging is biologically inevitable, see the article by Steven Austad listed in Suggested Readings.

19. S. Stearns, "Life History Tactics: A Review of the Ideas" *Quarterly Review of Biology*, 1976, 51: 3-37.
20. George Zug, *Herpetology: An Introductory Biology of Amphibians and Reptiles* (San Diego: Academic Press, 1993) 189.
21. Russel Bonduriansky, "A New Neartic Species of *Protopiophila* with Notes on its Behaviour and Comparison with *P. Latipes*," *The Canadian Entomologist*, 1995, 127: 859-863.
22. G. Bubenik, and A. Bubenik, *Horns, Pronghorns and Antlers* (New York: Springer-Verlag, 1990) 3-113.
23. Bonduriansky, "Storm in a Teacup," *Seasons*, Summer 1996: 27-31.
24. James Harding, and T. Bloomer, "The Wood Turtle, A Natural History," *HERP, Bulletin of the New York Herpetological Society*, 1979, 15: 9-26.
25. D. Foscarini, *Demography of the Wood Turtle and Habitat Selection in the Maitland River Valley* (University of Guelph M.Sc. Thesis, 1994).
26. Norman Quinn and Douglas Tate, "Seasonal Movements and Habitat of Wood Turtles in Algonquin Park, Canada, *Journal of Herpetology*, 1991, 25 (2): 217-220.
27. Quinn, "Turtle Trouble," *Nature Canada*, 1991, 20 (4): 21-25
28. The biology of the fly is poorly understood but is being investigated by researchers in Quebec.

## Chapter 11: The Station

1. Jack Korfield, *Bhudda's Little Instruction Book* (New York: Bantam, 1994) 152.
2. The fly diversity work has not yet been published but the results can be accessed at the Web site of the Department of Environmental Biology at the University of Guelph, or by e-mailing smarshal@evb-hort.uoguelph.ca.
3. D.A. MacLulich, "Fluctuations in the Numbers of the Varying Hare," *University of Toronto Studies, Biological Series*, 1937, 43. 136 pp.
4. C.H.D. Clarke, "Fluctuations in Numbers of Ruffed Grouse," *University of Toronto Studies, Biological Series*, 1936, 41. 118 pp.
5. Most of the material used here in describing the early history of the Wildlife Station comes from an interview with Bruce Stephenson on Dec. 15, 2000, in Gravenhurst, Ontario.
6. Jim Lowther and Dan Wood, "Specificity of the Black Fly, *Simulium euryadminiculum*, Toward its Host, the Common Loon," *The Canadian Entomologist*, 1964, 96: 911-913. The description of the study relayed here comes entirely from this fascinating paper.
7. Ibid, 912.
8. Stan Vasilauskas, Ontario Ministry of Natural Resources, Personal Communication, Fall 1999.

## Appendix I: Technical Expansion on Selected Topics

### A. Wolf Predation on Moose

1. Lloyd Keith, "Population Dynamics of Mammals," *International Congress of Game Biologists*, Stockholm, Sweden, 1973, 2-58.
2. P. Jordan, et al., "Biomass Dynamics in a Moose Population," *Ecology*, 1971, 52(1): 147-152.
3. Keith, 51.
4. Ian Thompson and Randolph Peterson, "Does Wolf Predation Limit the Moose Population in Pukaskwa Park?: A Comment," *Journal of Wildlife Management*, 1988, 52: 556-559. Anthony Bergerud and J. Barry Snider, "Predation in the Dynamics of Moose Populations: A Reply," *Journal of Wildlife Management*, 1988, 52(3): 559-564.

### B. Stocking and Genes

1. Peter Ihssen, "Genetic Diversity of Algonquin Trout Populations: Influence of Non-Native Genes," Ontario Ministry of Natural Resources, Unpublished Report.
2. Ibid.

### C. Carotenes

1. Tim Beardsley, "The A Team," *Scientific American*, 1991, February: 16-19.
2. Ibid. The question of the beneficial effects of beta carotene and retinoids is very much in dispute.

### D. Stress

1. The description of the stress response related here can be found in any general text on biology. For a more in-depth account, see A. Guyton, *Textbook of Medical Physiology*. New York: Saunders, 1981. See also *The Globe and Mail*'s Web page article of Tuesday, Dec. 21, 1999, at www.globeandmail.ca.

### E. Density Dependence

1. There is a rich literature on this subject. For one good example, see A. Sinclair, "Population Regulation in Animals" in J. Cherrett, *Ecological Concepts* (Oxford: Blackwell Scientific, 1989) 197-241.

# Bibliography

## For the General Reader

Paul Errington, *Of Predation and Life*. Ames: Iowa State University Press, 1967.

Aldo Leopold, *A Sand County Almanac*. New York: Ballantine, 1966.

Dan Strickland and Russ Rutter, *The Best of the Raven*. Whitney: The Friends of Algonquin, 1996.

Michael Runtz, *Moose Country*. Toronto: Stoddart, 1991.

## For Those with a Basic Background in Biology

Steven Austad, "On the Nature of Aging." *Natural History*, 1992 (2): 25-51.

James Bailey, William Elder, and Ted McKinney, *Readings in Wildlife Conservation*. Washington: The Wildlife Society, 1974.

Albert Franzmann and Charles Schwartz, *Ecology and Management of the North American Moose*. Washington: The Smithsonian Institute Press, 1998.

Stephen Herrero, *Bear Attacks: Their Causes and Avoidance*. Piscataway: Winchester Press, 1985.

Douglas Pimlott, J. Shannon and George Kolenosky, *Ecology of the Timber Wolf in Algonquin Provincial Park*. Toronto: Ontario Ministry of Natural Resources, Fish and Wildlife Research Branch Report No. 87, 1977.

Jonathan Weiner, *The Beak of the Finch*. New York: Knopf, 1994.

Bill Willers, *Trout Biology: A Natural History of Trout and Salmon*. New York: Lyons and Burford, 1982.

## For Those With Great Patience

H.G. Andrewarth and L.C. Birch, *The Distribution and Abundance of Animals*. Chicago: University of Chicago Press, 1974.

Paul Errington, *Muskrat Populations*. Ames: Iowa State University Press, 1963.

# Index

Acacia Trees, 148
Addison, Ed, 17-20, 52, 161, 162, 184, 207, 223
Adrenaline, 217, 218
Africa, 85, 89, 90, 135, 157
African Eye Worm, 7
African Savannah, 148
Alaska, 49, 87, 93, 146, 178
Albert Einstein Medical Centre, 124
Alberta, 129, 139
*Alces alces Americana*, 145
*Alces latifrons*, 145
Algonquin Park: vi, vii, 3-9, 13, 14, 16, 17, 19, 23, 25-27, 29-40, 44-46, 48, 55, 56-61, 63, 66, 67, 70, 71, 74-76, 79, 82, 83, 86, 87, 90-95, 99-105, 108-110, 112-115, 117-120, 123, 125, 128-130, 137-141, 144, 145, 149, 154-156, 158, 159, 161, 163, 164, 167, 169, 171, 173, 174, 176, 178, 179, 183, 185-187, 190, 192, 195, 196, 198, 199, 201-203, 205, 206, 209, 211-213, 215, 216, 224
  abundance of deer, 36, 37
  bass in, 117, 119
  description of, 4
  endangered, at risk, species in, 100, 198, 201, 202
  fishing in, 103, 104, 117, 119
  fish stocking in, 108-110, 114
  forest fires, 154
  founded in, 4, 23
  glaciation of, 34, 102
  hemlock failing, 38, 39
  hunting wolves, 58
  logging in, 29-32, 37, 74, 102, 154
  moose population in, 19, 37, 64

  songbird study, 45, 46
  summer weather in, 196
  trout fisheries, 104
  winters in, 14, 35
  wolf poisoning, 59, 82, 83
  wolves in, 56-60, 62, 128
American Academy of Family Physicians, 122
American Association for the Advancement of Science, 50
*American Zoologist*, 74
Ames (Iowa), 49
Anderson, Roy C., 7-10, 20, 36, 37, 149, 166, 178, 207, 211
Andrewartha, Herbert, 3, 5
Anthocyanins, 115, 116
Antiherbivory, 148
Antler Fly, 196, 197
Arnason, J.A., 111
Asio Lake, 105
Ataxia (paralysis), 127
Australia, 123
Australopithecenes, 89

Bacteria, 125, 141, 142, 147
Bailey, Ralph, 125
Baffin Island, 70, 101
Bald Eagle, 85
Bancroft (ON), 311
Banff National Park, 23
Bartlett, George, 55
Bass: 102, 117, 119
  Smallmouth, 117
Bateman, Robert (Bob), 208
Bates Island, 94
*Beak of the Finch*, 104
Bear(s): 71-73, 86, 91-100, 119, 150, 155, 156, 169
  Black, 71-73, 91-99, 155, 156, 162
  Grizzly, 73, 96-99, 155
  Polar, 97
Bear attacks, 94-99, 162, 226

Beaver(s), 59, 64, 75, 149-155, 157, 158, 180, 207
Beckwith township, 24
Beech: 25, 101, 156, 201
  American, 155
Bergereud, A.T. (Tony), 77-83, 92, 93, 194, 209, 214, 215
Beta carotene, 115-16, 216-17, 236
Bet hedging, 195, 209
Bice, Ralph, 33, 34, 71, 224
Big Trout Lake, 95
Biodiversity, 108, 204
Bird, David, 27
Birch, Louis, 3, 5
Birch, Yellow, 32, 207
Blackflies, 136, 137, 139, 158, 190, 197, 204, 209-210
Blumenshine, Robert, 88-90
Bombardier J5, 207
Bond, C.J., 58
Bondurianksy, Russell, 196, 197, 212
Bonnechere River, 29
Booth, John Rudolphus, 28
Brainworm (pitenuis), 10-12, 16, 20, 36, 37, 41, 126, 127, 141, 178, 211
British Columbia, 105
Britton, Don (Dr.), 26
Brooks, Ron (Prof.), 187, 188-197, 199, 201, 211
Brunton, Don, 26
Bubenik, Anthony B. (Tony) (Dr.), 163-177, 185, 212, 232, 233
Buss, Mike, 164, 177, 232, 233

Cache Lake, 5, 36, 144
California, 49, 115, 148
Cameron Lake, 167
Canadian Aviation Hall of Fame, 63
Canadian Shield, 32
Caribou, 32-35, 70, 76-78, 81, 91, 101, 173, 194, 209, 214
Carotenoids, 115-117, 216
Carrion Crows (Europe), 157

Cartier, Jacques, 32, 40
Cat Scratch Fever, 52, 53, 92
Cavallo, John, 88-90
Cellulose, 141
Champlain, Samuel de, 33, 40
Chapleau (ON), 64
*Chelydra serpentina*, 186
Chicago (IL), 133
Christian, John, J., 124, 125, 134, 135, 218
Christie, W.J., 119
Clarke, C.H.D. (Doug), 205, 206
Cloquet (MN), 54
Code, Ted, 129, 130
Cole, Bill, 41-44, 178
Columbia River, 106
Comfort, Alex, 194
Commensalism, 149
Compensation, 54, 56, 59
Cornell University, 120
Cornish, Matt, 6
Cortisol, 134, 135, 217, 218
Costello Lake, 111
*Corvidae*, 157, 179
Coyote, 100
Cro Magnon, 88
Crowe Lake, 17
Crusaders, 44
Czechoslovakia, 163


DDT, 202
Danzmann, Roy, 109, 215
Darwin, Charles, 104, 178, 188
Deer:
  Mule, 34
  Red, 173
  White-tailed, vii, 9-13, 20, 21, 32-41, 56, 57, 59, 65, 66-70, 73, 77, 82, 83, 84, 93, 102, 112, 119, 120, 126, 127, 136, 141, 143, 144, 147, 153, 154, 167, 170, 207, 214, 224

bear predation, 92, 93
brainworm in, 9-13
census of, 65, 66
eating Hemlock, 36, 40
history in Park, 32-35
rumenitis in, 143
snowmobiles, 120
winter starvation, 35, 37, 74, 120, 136, 143
wolf-killed, 66-70, 73, 76
Dendroica Lake, 105
Dennison, John (Capt.), 94
Density dependence, 218
*Dermacentor albipictus*, 14
Desser, Sherwin, 207
Detroit (MI), 130, 131
*Distribution and Abundance of Animals*, The, 3
Dogs, Wild (Africa), 135
Dolphin, 81
Doomed surplus, 53, 54, 56, 93

Earthworms, 201
Ebola (of Africa), 51
Eagle, Golden, 83
*Echinococcus granulosus* (see Tapeworm)
Elephants, 192, 195
Elk, 33-35, 170
England, 30, 43
Errington, Paul, 2, 47-54, 56, 65, 73, 82, 99, 100, 214
Espanola (ON), 92
Ethologists, 161
Eucalia Lake, 82
Evolution, 188, 189

Falls, J. Bruce, 156, 157, 158, 206, 207
Finland, 82
Fisker, Yorkie, 63
Fly, flies, 204, 205, 209, 235
Florida, 95, 183

Florida (Everglades), 35
Forbes, Graham, 87, 88
Found Lake, 101
Fox(es), 82 86, 192
France, 3
Fraser, Jim, 110, 112, 113, 212
Frog(s), 123, 124, 138, 187, 190
Fryxell, John, 149-152, 156, 158

Galapagos Islands, 104
Gervais, Jack, 58
Giraffe(s), 148
Glucose, 217, 218
Goshawks, 54
Graham, Sam A., 26
Grant, Peter, 104
Grant, Rosemary, 104
Gravenhurst (ON), 207
Great Gray Owl, 201
Grebe(s), 209, 210
Greenleaf Lake, 26
Grey Islands, 80
Grouse: 155
    Ruffed, 54, 136, 138, 205
    Spruce, vi, 138, 139, 159, 160
Gullion, Gordon, 54, 82
Gull(s), 210
Gut flora, 141
Guyette, Richard, 41-44, 178

Hacking, 202
Haliburton (ON), 187
Hall, Alex, 75
Hanford Reach, 107
Hare:
    Arctic, 79-82
    Snowshoe, 78-81, 93, 134, 139, 183, 205, 225
Harkness, ___ (Mrs.), 11
Harkness, William, 111
Harkness Laboratory of Fisheries Research, 42, 61, 62, 110, 111, 206

Harvard University, 29
Hemlock, Eastern, 25, 29, 32, 36-41, 209, 213, 222
Heron, Great Blue, 84, 138
Herrero, Stephen, 96, 97
Hess, Quimby, 4
Hibernacula, 194
Hognose Snake, 201
Holland, 148
*Homo Sapiens*, 88
Huntsman, A.G. (Dr.), 111
Huntsman, ___ (Mrs.), 111
Huronia, 35
Hydatid cysts, 21
Hydatid disease, 21, 75
Hyena(s), 85, 89, 90

Ihssen, Peter, 109, 114, 215, 216
Illinois, 182
Immune system, 217
Indiana, 132
Intermediate host, 12, 14
Iowa, 48, 49
Ireland, 203
Isle Royale (Michigan), 214, 215
Isoflavones, 148

Jacobson's organ (moose), 167
James Bay (Lowlands), vi, 64
Jay(s):
    Blue, 86
    Grey, vi, 179-185
    Scrub, 183, 184
Japan, 9, 10
Japanese scientists, 9, 10
Jenkins, Ted, 63
Jerusalem, 44
Joe Lake, 55
Joe Lake Station, 55

K-selected, 195, 201
Keddy, Paul, 26
Keith, Lloyd, 214, 215

Kennedy, W.A., 111
Keppie, Dan, 80
Keystone species, 154
Kingscote Lake, 99
Kirksville (Missouri), 133
Kolenosky, George, 75
Kreb's cycle, 202
Kujala, Henry, 212

La Verendrye Park, 179
Lactic Acid, 127
Lake Erie, vi
Lake Huron, 211
Lake of Two Rivers, vi, 61
Lake Opeongo, 61, 63, 94, 111, 119, 206
Lake Sasajewun, vi, 208
Lake Superior, 214
Lakehead University, 11, 166
Lankester, Murray (Prof.), 11, 166
Lascaux paintings, 3, 5, 88, 219
Leopards, 89, 90
Leopold, Aldo, 2, 5, 137, 148, 219
Leopold, A. Starker, 148
Lignin, 141
Lion(s), 85, 88
Llewellyn, Linda, 27
*Loa loa* (see African Eye Worm)
Lone Creek, 94
Loon(s), 8, 59, 110, 112. 158, 209, 210
Louisiana, 49
Lowers, Arthur, 29
Lowther, Jim, 209, 210
"Lucy," 88-90
Lynx, 52, 77-81, 83, 92, 93, 134
*Lumberjacks, The*, 28
Lymphocytes, 10

MNR (Ministry of Natural Resources), 26
MacDougall, Frank, 63, 206
MacLulich, D., 205

Madawaska Valley, 31
Maine, 23, 34, 35, 198
Manitoba, 35, 156
Maple: 25, 101, 201
    Sugar, 115, 147
Maple (ON), 17
Marquette (MI), 125
Marshall, Steve, 204
Marten(s): 82, 121, 122
    Pine, 95
Martin, Nick, 94
Martin, Norman Duncan (Norm), 45
Maryland, 23, 198
McCoy, Karen, 13
McIsaac, Donald, 106-108
McLaughlin, Rob, 8
Mech, David, 72
Meningitis, 11
Merganser(s), 210
Michigan, 121, 122, 125, 126, 132, 145, 198, 214
Michigan Department of National Resources, 125
Michigan moose transfer, 125, 126, 145
Middlesex County, 34
Millar, John W. (Jack), 36, 58
Minnesota, 54, 72
Missanabie (ON), 92
Missouri, 23, 34, 101, 128, 129-131, 133, 134, 213
Missouri Department of National Resources, 133
Moen, Aaron, 120
Mohawk First Nation, 33
Mollusks, 117
Monkeys, Red Colobus, 157
Montana, 73
Montreal, 1
Moose: vi, 8-11, 13-21, 32-37, 40, 41, 57, 62, 64, 71, 76, 77, 86-88, 90-93, 102, 119, 125-127, 139-147, 149, 153, 154, 159, 160-178, 184, 196, 197, 209, 211, 212, 214, 215, 217, 234
    antler locking, 178
    antlers, 145, 167-171, 176-178, 196, 197, 212
    attacks, 169, 178
    behaviour, 164, 167
    bell(s), 165, 166
    brainworm, 9-11, 13-16, 20
    calf, calves, 8, 18, 91, 92, 141-143, 159, 160, 169, 174
    description, 8, 14, 145
    diet of wolves, 57, 75, 87, 88, 90, 91, 214
    face colour, 176
    fighting, 171, 176-178
    history of, 32-34
    hooves, 164
    Michigan reintroduction, 125, 126
    milk of, 141, 142
    nose, 167, 168
    numbers in Park, 19, 32, 37
    nutrition of, 139-144, 146, 147
    penis, 168
    predation by bears, 93
    rut, rutting, 8, 142, 143, 167-169, 171, 172, 175, 176, 178, 234
    salt and roads, 139, 140
    stomach, 142
    surveys, 71
    swimming, 146
    temperature stress, 140, 141
    territoriality in, 178
    ticks, 14-18
    weights of, 144-146
Mosquito Bomber, 31
Musk Ox(en), 101
*Muskrat Populations*, 48, 51
Muskrat(s), 47, 49-53, 56, 57, 65, 214
Musky (muskellunge), 102
Myopathy, 127

Nancekivell, Graham, 191

Native North Americans (Native cultures), 3, 23, 25, 32, 103, 114
Necrotizing Facsiitis, 124
Neotropical migrants, 44
New Brunswick, 20, 159
New England, 32, 34
New France, 33
New York, 203
Newfoundland, 49, 57, 76-81, 144, 209, 205, 225
Nichols, G.E., 26
North Bay (ON), 33
Northern Hardwood Forests, 24, 27
Nova Scotia, 9
Nudds, Tom, 13

Obbard, Marty, 192
Ocam's Razor, 95
Odalisques, 169
*Of Predation and Life*, 99
Ontario Fisheries Research Laboratories, 111
Osprey, 84
Ottawa (ON), 28, 35
Ottawa First Nation, 33
Ottawa River, 29
Otter Transfer, 128-134
Otter(s): 83, 84, 86, 112, 128-134, 193, 195, 209, 211
    River, 128
Ovenbird(s), 156, 157

Panama, 123
Parasitology, 7
*Parelophostrongylustenuis* (see brainworm)
Parry Sound District, 66
*Pasturella multicoda*, 77
Peel, Paul, 219
Pennsylvania, 41, 109, 198, 200
Peregrine Falcons, 202, 203
Perych, Peter, 27
Petawawa River, 29

Peterson, Randolph, 145
Peueria, 148
Phenols, 147
Philadelphia (PA), 123
Phillips, George, 63
Pimlott, Douglas (Doug), 57-61, 63, 66-69, 71, 73-75, 82, 83, 87, 149, 207, 209, 224
Pine(s): vi, 131, 201
    Jack, 138, 139, 160
    Red, 29
    White, 25, 27, 29, 37, 38, 43
Pneumonia, 15, 77, 91
Poland, 123, 164
Pollen, 27, 40, 101
Polygenes, 114
Predator pit, 80
Prey switching, 76, 79
*Protopiophila litigata*, 196
Pruit, Bill, 140, 230

Quail, 148, 214
Quebec, 179, 237
Quebec City (QC), 29
Quinn, Norm, vii
Quinn, Tom, 106-108

R-selected, 195
Rabbit(s), 192
Rabbit Fever, 89
Rabies, 61
Raccoon(s), 83, 100, 192, 195, 199
Radio telemetry, 59, 135
Rainforest, 45
Rae, Bob, 34
Raven(s), 83, 86, 90, 156
Raven Creek, 91
Red Wolf (see Wolf, Red)
Redds, 112
Redrock Lake, 13
Releaser(s), 161, 168, 169, 171, 176, 177
Retinoids, 216, 217, 238

Reznucek, Tony, 26
Rhode Island, 106
Ridgway, Mark, 42, 43
Robb, D.L., 111
Rock Lake, 55, 118
Royal Ontario Museum, 145
Ruffed Grouse, 54, 136, 138, 205
Rumenitis, 143
Rumination, 143

Sagadahoc River, 32
Salamanders: 42
    Red-backed, 42
    Yellow-spotted, 42
Salmon: 114
    Chinook, 106, 107
    Pacific, 106, 107
Salmonella, 124
*Sard Country Almanac, A*, 2, 148
Saskatchewan, 35, 93
Scientific American, 88
Scott Lake, 41, 42, 204, 205
Scutes, 192, 204
Serengeti National Park, 90
Seton, Ernest Thompson, 35, 156
Shannon, Jack, 69, 70
Sierra Nevadas, 115
*Simulium euryadminiculum*, 209, 210
Skunk(s), 100, 132, 192, 199
Smith, Andrew, 45, 46
Smith, Don, 208
Smoke Lake, 67, 206
Smith, Peter, 95, 96, 167, 173, 175, 176, 224
Snake:
    Eastern Hognose, 201
"Social Subordination, Population Density, and Mammalian Evolution," 125
South Dakota, 47
Squirrels:
    Red, 183
St. John's (NF), 79, 144

St. Lawrence Valley, 32
Stanfield, Rod, 207
Stephenson, Bruce, 58, 207, 224, 235
Stickleback(s), 105, 106
Stocking of fish, 102, 108-110, 112, 114, 215
Stomping (wood turtle), 200
Standard Stress Model, 120
Stress response: 122-125, 131, 134, 217, 228, 229, 238
    Capture stress, 124
Strickland, Dan: vii, 26, 176, 178-185, 233
Submerged wood, 42
Suckers, 216
Supplementary feeding, 143, 144
Swan Lake Forest Reserve, 41-44, 207

Tanzania, 90, 135
Tapeworms (E. granulosus), 21
Taylor, Jim, 31
Taylor, Henry, 31
Terpenes, 147
Theberge, John, 87, 88
Thomas, Edwin, 55
Thomas, ___ (Mrs.), 55
Thomson, Tom, 103
Thoreau, Henry David, 24
Thunder Bay (ON), 11
Tick (see Winter Tick)
Timmerman, Tim, 165, 166
Toronto (ON), 32, 37, 52, 55, 58, 71, 129, 132
Trout: 102-105, 109, 112, 114-117, 108-112, 114-117, 215, 216, 227
    Aurora, 110
    Brook, 42, 83, 84 102-106, 108-112, 114-117, 194, 195, 215, 228
    Golden, 115
    Lake, 102-104, 118, 213

Salmon, 118
"Silver," 114, 115, 228
Tudor, Mike, 196
Tulip(s), tulip bulbs, 148
Turkey, Wild, 125, 128
Turquoise Lake, 71
Turtles: 185, 190-196, 209, 234
    European Pond, 195
    Painted, 190
    Sea, 194
    Snapping, 53, 186-196, 201, 211
    Wood, 100, 198-201
Turvi, Harry, 94

U.S. Fish and Wildlife Service, 203
Umwelt, 166
University of Calgary, 96
University of Guelph, 7, 8, 10, 13, 26, 149, 187, 188, 204, 237
University of Manitoba, 140
University of Minnesota, 54
University of New Brunswick, 80, 160
University of Toronto, 45, 119, 123, 156
University of Waterloo, 87
University of Wisconsin, 214, 219
Upper Canada, 24
Upper New York State, 121
Upper Peninsula (Michigan), 125, 126, 214

Vasiliauskas, Stanley (Stan), 38, 39, 178, 213
Vikings, 43
Voigt, Dennis, 74-76, 87
Vole(s): 49, 121, 125, 135
    Red-backed, 121
Vulture(s), 85, 86, 90

Warbler(s), 105, 137-139, 148, 158
Washington State Department of Fisheries, 106
Weatherwax, Wendell, 27

West Africa, 7
Western Uplands (Algonquin Park), 113
Westward Lake, 103
Whiskey-jack(s) (see Grey Jay)
White blood cells, 10, 124, 218
Whitefish, 102
Whitefish Lake, 202, 203
White Trout Lake, 202
Whitney, G.G., 29
Whitney (ON), 30
Whitson Lake, 34
Wild Turkeys, 125, 128
Wildlife Research Station, vi, 6, 19, 58, 62, 122, 128, 154, 156, 158, 189, 205, 206, 208, 209, 211, 235
Wilton, Mike, 86, 92, 93
Winterkill, 193
Winter Tick, 14-20, 184, 213
Wisconsin, 23, 194
Wisconsin Ice Sheet, 101
Wolf, wolves, vi, 21, 37, 39, 56-62, 66-76, 79, 82-84, 86-88, 90, 91, 100, 112, 127, 128, 161, 165, 169, 209, 214, 215, 224, 225
    Arctic, 101
    Grey, 57
    hunting skills, 66-68, 74
    Red, 57
Wolf, Roger, 27
Wood, Dan, 209, 210
Woodcock, 3
World War II, 2, 32, 57, 148, 163

Year class strength, 119
Yellowstone Park, 23
Young, David, 17
Young, Tim, 17
Yukon, 134, 135

Zach, Reto, 156-158

## About the Author

Norm Quinn has been Park Management Biologist in Algonquin Park, Ontario, since 1984. He received his B.Sc. in wildlife management from the University of Guelph in 1976 and his M.Sc. in wildlife biology from the University of New Brunswick in 1978. Norm has had an extraordinarily varied career in fish and wildlife research and management, having worked with everything from minnows to moose. He has published widely in the popular and scientific literature and has travelled extensively, having lived for lengthy periods in both Israel and Australia.

Norm Quinn lives in Bancroft, Ontario, with his wife Nancy, son Robert, and daughter Laura.

### About the Illustrator

Cassandra Ward, born in Bancroft, Ontario, in 1983, has always had a passion for art. Her aspirations and commitment stem from the ongoing encouragement of her family. In June of 2000 she was presented with the E.W. Smith Award, and she has won a number of awards at the annual Student Art Exhibitions in Bancroft. Having recently graduated from North Hastings High School, Cassandra will begin attending the University of Waterloo in September 2002, aiming for a general Arts degree with honours.